USSR IN MAPS

USSR IN MAPS

J. C. Dewdney MA
Reader in Geography
University of Durham

DISTRIBUTED BY
HOLMES & MEIER
PUBLISHERS, INC.
New York   London

ISBN 0 340 24414 3 ✓

First published 1982

Typeset by Macmillan India Ltd, Bangalore
Printed in Hong Kong for
Hodder and Stoughton Educational,
a division of Hodder and Stoughton Ltd,
Mill Road, Dunton Green, Sevenoaks, Kent
by Colorcraft Ltd

CONTENTS

Introduction vi

Part 1 *Physical Environment* 1
1 Size and position 2
2 Geology 4
3 Relief 6
4 Drainage 8
5 and 6 Climate 10
7 Soils 14
8 Vegetation 16
9 Natural regions 18

Part 2 *Human Geography* 21
10 Political divisions 22
11 Economic and administrative divisions 24
12 Growth of the state 26
13 and 14 Population distribution and density 28
15 Population change, 1970–9 32
16 Nationalities 34
17 Towns 36

Part 3 *Economic Geography* 39
18 and 19 Crop distributions 40
20 Agricultural regions 44
21 Agricultural land use and production 46
22 Minerals 48
23 Coal 50
24 Oil and natural gas 52

25 Electricity 54
26 Iron and steel 56
27 Non-ferrous metallurgy 58
28 and 29 The engineering industries 60
30 and 31 The chemical industries 64
32 and 33 Textile, clothing and footwear industries 68
34 Timber industries 72
35 Alimentary industries 74
36 Railways 76
37 Roads 78
38 Waterways 80
39 Foreign trade 82

Part 4 *Regions* 85
40 Key to the regional maps 86
41 The North-West, Baltic and Belorussian regions 88
42 The Centre, Volga-Vyatka and Black Earth Centre regions 90
43 The Ukraine and Moldavia 92
44 The Volga region 94
45 The Ural region 96
46 Siberia and the Far East 98
47 The North Caucasus and Transcaucasia 100
48 Kazakhstan 102
49 Central Asia 104

Bibliography 107
Index 109

INTRODUCTION

This volume contains 49 maps and diagrams, each with a page (or more) of supporting text, designed to illustrate the present-day geography – physical, human and economic – of the Union of Soviet Socialist Republics, the world's largest political unit. In view of the chosen format, in which a single page of text accompanies each map page, the written analysis is inevitably brief and summary; much fuller information is available from the works listed in the bibliography. The relatively generous page size of the present volume permits the production of maps on a much larger scale and with considerably more detail than in most standard geographies of the Soviet Union and *USSR in Maps* is intended to supplement rather than to compete with such books. Nevertheless, it is hoped that the textual material in this volume will provide a wide range of basic information on the geography of the Soviet Union and thus fill the role at least of an introductory survey.

As befits its title, it is the maps rather than the text, which are unique to this volume and thus some discussion of their nature seems appropriate. Despite the dimensions of *USSR in Maps*, cartographic representation of the Soviet Union presents serious problems which result from the country's basic geographical features of size and shape. On a page 230 mm × 270 mm, the largest scale at which it is possible to print a rectangular block enclosing the whole territory of the USSR is approximately 1 : 33,000,000 (1 cm = 330 km or 1″ = 520 miles), which may be compared with the case of *Sierra Leone in Maps*, the first volume in this series, where it was possible to map the whole country within a square 180 mm × 180 mm at a standard scale of 1 : 1,900,800 (1 cm = 19 km or 1″ = 30 miles), 17 times as large as that possible for the USSR. Obviously enough, the amount of detail which can be shown is a function of the size of the territory involved as well as the amount of information available.

The problem of size is exacerbated by that of shape. A rectangle precisely enclosing the USSR at a scale of 1 : 33 million measures 140 × 250 mm, a ratio of 1.8 to 1, and the fact that the USSR covers a much greater distance east-west than north-south further restricts the map scale. However, it also means that the 230 × 270 mm format provides extra space at the bottom of each page for key, inset maps and/or other types of diagram, and full advantage has been taken of this wherever possible.

Further problems arise from the extremely uneven distribution of population and of most forms of economic activity within the confines of the Soviet Union, a point referred to in many of the texts. As extreme examples, we may quote the Moscow oblast, which covers 47,000 sq. km and contains about 12,800,000 people, and the Yakut ASSR with an area of 3,103,200 sq. km and a population of only 664,000; while the former contains nearly 200 urban places, including 15 cities with more than 100,000 inhabitants each, the latter has only about 60, of which only a dozen are fully-fledged towns and only one (Yakutsk : 152,000) exceeds the 100,000 threshold. Industrial and other activities of all types show similar degrees of concentration and dispersion which makes it virtually impossible in many cases to display a single feature by a single map covering the entire country.

Consequently, the idea of a set of maps to a common scale and drawn on a common base covering the whole of the USSR was rejected at the outset on practical grounds and a variety of formats have been adopted to permit greater flexibility and a choice of scales. Nearly half the 38 'national' maps (i.e. excluding those depicting specific regions) do in fact show the whole of the Soviet Union on a common scale, for example those of aspects of the physical environment (Maps 2: *Geology*; 3: *Relief*; 4: *Drainage*; 7: *Soils*; 8: *Vegetation*); maps showing various kinds of regionalisation (9: *Natural Regions*; 10: *Political Divisions*; 11: *Economic and Administrative Divisions*; 20: *Agricultural Regions*); some of the economic maps (23: *Coal*); and the maps showing population (14, 16) and communications (36, 37, 38).

Elsewhere, a variety of strategems have been adopted to provide the maximum amount of information in the space available. In many cases, this has involved the use of a more generous scale for the densely settled western regions, with parts of Siberia and the Far East reduced to smaller-scale insets (e.g. 13: *Population Distribution*; 17: *Towns*; 25: *Electricity*; 26: *Iron and Steel*); large-scale insets of areas of special complexity, such as the Moscow basin, Donbass or Urals also appear in several places. In yet other cases, topics too complex to be covered satisfactorily by a single black and white map are displayed by several maps on the same page as, for example, those of *Climate* (5, 6); *Crops* (18, 19) and *Minerals* (22). In several instances the two techniques are combined, notably in the standardised set of maps showing various types of industrial activity (28–35). The most complex layout is that of Map 16: *Nationalities*, which is designed to show the distribution of more than 50 ethnic groups.

The topics themselves follow, it is hoped, a fairly logical sequence. Following Map 1 (*Position*) where the Soviet Union's world location is shown from two different angles, Maps 2–8 cover various aspects of the physical environment, which combine to produce the *Natural Regions* of Map 9. Maps 10 and 11 show man-made political, economic and administrative divisions of the massive territory amassed under a single administration over the past 400 years (Map 12: *Growth of the State*). Various aspects of population geography are illustrated by Maps 13–17. Economic aspects are portrayed in about 50 maps and diagrams arranged on 18 pages (Maps 18–35), followed by four pages covering transport and trade (36–39). Finally, there is a set of nine regional maps (41–49) the layout of which is discussed in the text to Map 40: *Key to Regional Maps*.

The mammoth task of preparing the artwork for this volume – there are well in excess of a hundred maps and diagrams in the book – would never have been completed without the skill and devotion of the Cartographic Section, Department of Geography, University of Durham. Nor could the text have been produced without the typing skills of the Department's secretarial staff. Formal expressions of gratitude cannot adequately convey my thanks to them all.

J. C. Dewdney

Durham
March, 1982

PART 1 PHYSICAL ENVIRONMENT

1 SIZE AND POSITION

With a total area of 22,402,200 sq. km, the Soviet Union is, by a wide margin, the world's largest political unit, more than twice the size of its nearest rivals Canada (9,976,000 sq. km), China (9,651,000 sq. km) and the USA (9,363,000 sq. km), 4.5 times the size of the rest of Europe (4,946,000 sq. km) and no less than 91 times that of the United Kingdom (245,000 sq. km). Soviet territory accounts for one-sixth of the earth's land surface and covers about 40 per cent of the Eurasian land mass, including nearly two-thirds of Europe and more than one-third of mainland Asia.

In terms of population, however, the USSR is somewhat less impressive. Her 1980 population of 265 million was only one-seventeenth of the world total (4,500 million), ranking third after those of China (1,042 million, est.) and India (684 million) and exceeding that of the United States (222 million) by less than 20 per cent. The Soviet population is about 4.7 times that of the United Kingdom (56 million). The relatively low average population density of the USSR (Map 14) results mainly from her position in high latitudes and the resultant predominance of hostile physical environments (Maps 5—9).

As Map 1A shows, the Soviet Union (excluding offshore islands) covers a latitudinal range of 43°, extending from 35°N in the extreme south of Soviet Central Asia to 78°N at Cape Chelyuskin in the Taymyr peninsula. Her territory widens northwards to a maximum along the Arctic Circle and more than three-quarters of the country is north of the 49th parallel (the northern boundary of the USA excluding Alaska). Leningrad, the most northerly of the Soviet Union's leading cities, lies on the 60th parallel, the same latitude as the southern tip of Greenland and Anchorage, Alaska, while Moscow (56°N) is in the same latitude as the centre of Hudson Bay.

While the Soviet Union's maximum north-south extent is about 4,000 km, the distance from Kaliningrad on the Baltic to the Bering Strait in the far north-east of Siberia exceeds 9,000 km and covers 170° of longitude from 20°E to 170°W, crossing eleven time zones. At noon GMT it is 3.00 p.m. in Moscow and 1.00 a.m. of the next day in the Chukot peninsula.

The boundaries of the USSR, along which she faces fourteen other countries, total some 53,000 km, of which about 38,000 km are maritime boundaries, mainly on the Pacific and Arctic Oceans with relatively short stretches along the Baltic, Black and Caspian seas. Of the 15,000 km of land boundary, about 3,000 km are in Europe, where the Soviet Union has political, frontiers with Norway (150 km), Finland (1,100 km) and her COMECON partners Poland (900 km), Czechoslovakia (100 km), Hungary (90 km) and Rumania (800 km). In the past, the greatest geopolitical significance was attached to this western frontier, but in recent years increasing attention has been paid to the southern boundary which stretches nearly 12,000 km from the Black Sea to the Pacific. Here the USSR abuts on Turkey (400 km), Iran (1450 km), Afghanistan (1800 km), China (5,400 km), the Mongolian People's Republic (2,600 km) and North Korea (50 km). In the far north-east, the USSR faces the USA (Alaska) across the 80 km wide Bering Strait, where the Russian Ratmanov (Big Diomede) Island and the US Little Diomede Island are a mere three kilometres apart. In the far south-east, only 10 km separate Yuzhno-Kurilsk, the most southerly of the Kurile Islands, in Soviet hands since 1945, from Hokkaido, Japan.

The conventional view of the world (Map 1A) shows the USSR separated from North America by western Europe and the broad expanse of the North Atlantic, but the polar view (Map 1B) demonstrates that only 1,500 km separate the most northerly islands of the USSR and Canada, whose mainlands are some 3,000 km apart across the Arctic. In the modern world of long-range aircraft and inter-continental ballistic missiles, the spatial relationships portrayed by the polar view are of greater significance than those of the traditional type of world map.

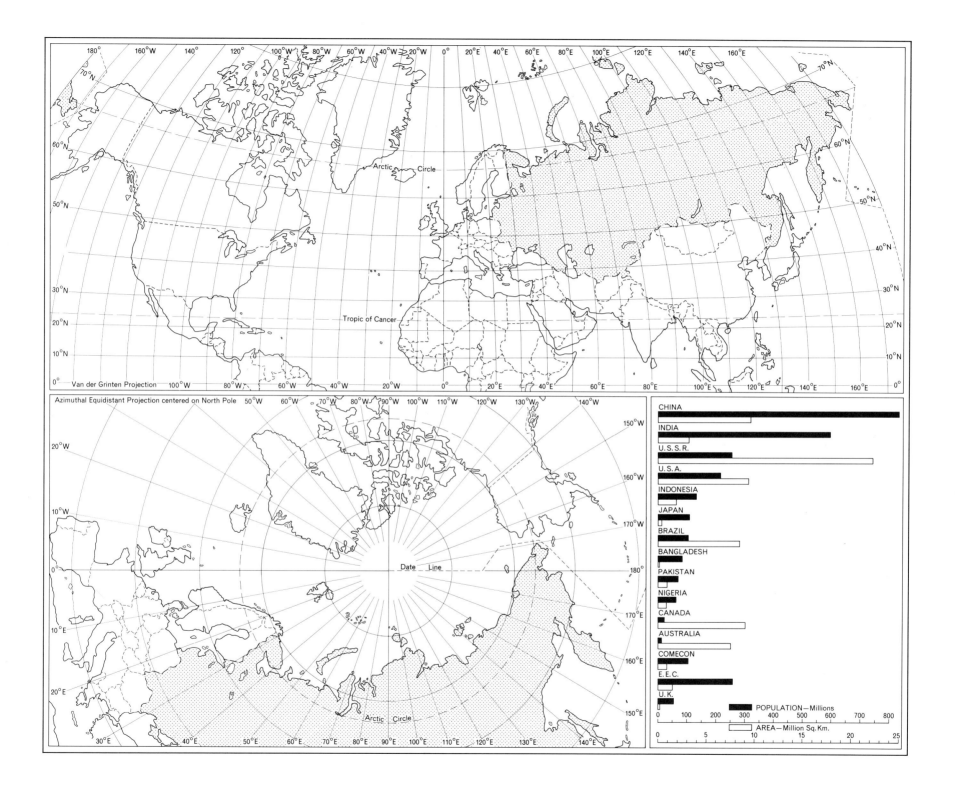

Van der Grinten Projection

Azimuthal Equidistant Projection centered on North Pole

Arctic Circle

Tropic of Cancer

Date Line

Arctic Circle

CHINA

INDIA

U.S.S.R.

U.S.A.

INDONESIA

JAPAN

BRAZIL

BANGLADESH

PAKISTAN

NIGERIA

CANADA

AUSTRALIA

COMECON

E.E.C.

U.K.

POPULATION—Millions

0   100   200   300   400   500   600   700   800

AREA—Million Sq. Km.

0   5   10   15   20   25

2 GEOLOGY

Not surprisingly, in view of its vast territorial extent, the Soviet Union contains sizeable areas of rocks of virtually every geological period and type. Even at the relatively generous scale used in this book, it is virtually impossible to show, in a single map, the full complexity of the geological structure of the USSR. Map 2 distinguishes only six categories of rocks, namely those derived from each of the five major geological eras – Pre-Cambrian, Paleozoic, Mesozoic, Tertiary and Quaternary – and intrusive materials.

The present-day distribution of these six categories can only be understood by reference to the geological history of the territory which now constitutes the USSR and by relating this to the structure and geological history of the world as a whole. Two major structural components are involved, namely stable blocks or 'continental platforms' and the intervening 'orogenic belts' or zones of mountain building. The continental platforms are composed of extremely tough igneous and metamorphic materials formed at depth during mountain-building episodes in early (Archaean or Pre-Cambrian) geological times and exposed by long periods of sub-aerial denudation. The platforms have resisted later folding movements but have been faulted and differentially raised and lowered by 'epeirogenic' earth movements. Thus the platforms are, in places, exposed at the surface but, for the most part, lie buried to varying depths beneath a cover of younger sedimentary rocks. The intervening orogenic belts are zones where, at various times, depressions in the earth's crust or 'geosynclines' developed, and received great thickness of sedimentary materials laid down around the older land masses. These sediments were subjected to folding in one or more of the great mountain-building episodes – Caledonian (mid-Paleozoic), Hercynian (late Paleozoic) and Alpine (late Mesozoic/Tertiary) – which gave rise to high ranges of fold mountains. Some of these – particularly from the Alpine orogeny – are major elements in the present-day landscape, but the mountain systems of the earlier Caledonian and Hercynian orogenies were subject to sub-aerial denudation and block faulting so that these structures, like those of the pre-Cambrian period, are partly exposed and partly concealed beneath younger deposits.

The Soviet Union contains two of the world's major continental platforms, the East European Platform between the Urals and the western boundary (which continues westward across the North European Plain to south-east England, where it is concealed beneath the sedimentaries of the Weald) and the Siberian platform, which occupies the territory between the Yenisey river and the fold mountains between the Lena valley and the Pacific.

Within the USSR, the intrusive and metamorphic materials of the East European platform are exposed in two areas. The larger of these, in the north, occupies Karelia and the Kola peninsula and is part of the area known as the Baltic or Fennoscandian 'shield', which also occupies Finland, most of Sweden and southern Norway. The smaller Ukrainian shield runs from north-west to south-east across that republic. Elsewhere, the platform is concealed beneath sedimentaries of varying age and surface rocks become younger towards the south and east. As Map 2 shows, Paleozoics occupy a triangular area from the Baltic and White Seas to the Moscow basin; there is a relatively narrow outcrop of Mesozoics to the south-west and a much larger one to the east, extending as far as the Ural Mountains. The surface rocks of most of the Ukraine are of Tertiary age, while Quaternaries occupy a depression running from the Sea of Azov to the Caspian and large areas to the north and north-east of the latter.

The Siberian platform is more complex and is also exposed in two areas – in the smaller, northerly Anabar shield and in the much larger Aldan shield to the east of Lake Baykal. On either side of the Anabar shield, in the Tunguska and Lena basins, the platform is deeply buried beneath Mesozoic and Paleozoic (including extensive Carboniferous) deposits; in the south, the platform was fractured by Caledonian and later orogenies to give large uplifted blocks and down-faulted basins.

Between the two platforms, a geosyncline developed during the Paleozoic and the great thicknesses of sedimentaries deposited therein were subjected to intense folding during the Hercynian orogeny, giving complex mountain systems. However, the Hercynian mountains then underwent a long period of denudation and were later faulted and experienced differential uplift. In the raised sections – the Ural Mountains and the Kazakh Upland – igneous and metamorphic materials, together with tightly folded Paleozoics, were exposed, but in the down-faulted sections – the West Siberian and Central Asian lowlands – Hercynian structures are deeply buried beneath younger sedimentaries and the surface rocks are wholly of Tertiary and Quaternary age.

Finally, in a zone running along the southern border of the country through the Crimea and Caucasus to Central Asia and over much of the territory to the east of Lake Baykal and the Lena, folding occurred on several occasions and Paleozoic, Mesozoic and Tertiary rocks all occur in close proximity. The most striking landforms are the product of the Alpine orogeny, which has given a complex system of fold mountain ranges.

The present-day relief, illustrated in Map 3, is the product of denudation processes operating over long periods on the complex geological structures and varied rock types outlined above.

GEOLOGY

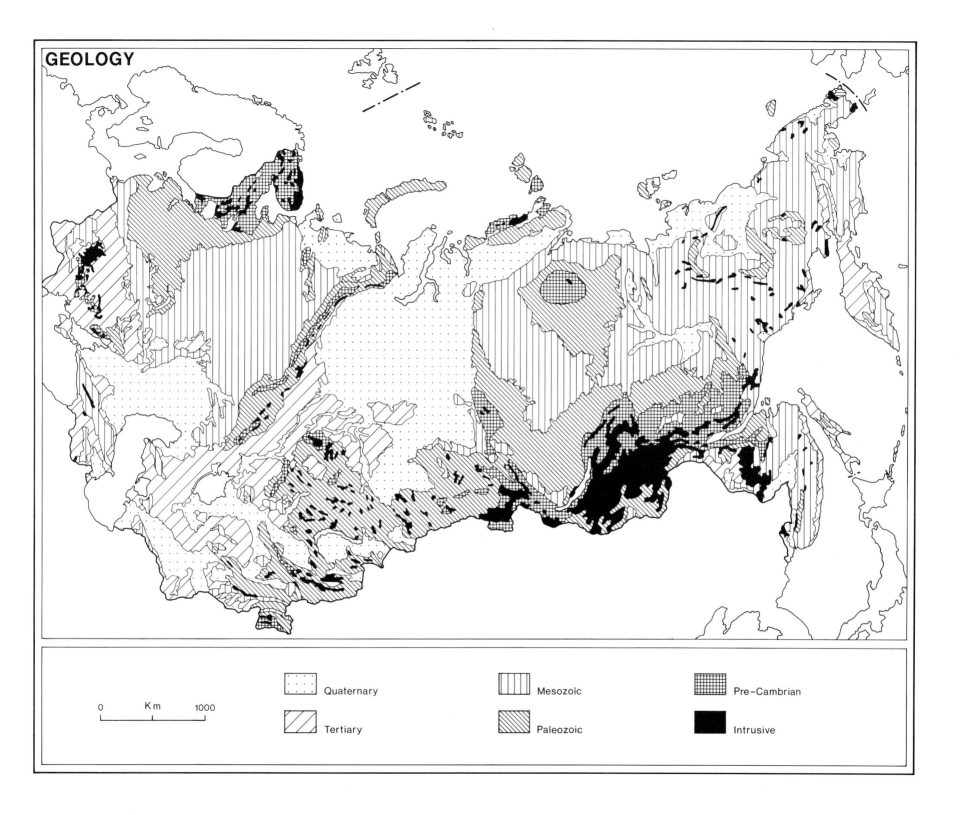

Quaternary	Mesozoic	Pre-Cambrian
Tertiary	Paleozoic	Intrusive

0   Km   1000

5

3 RELIEF

Despite the tremendous contrasts in relief and altitude which occur within her borders, the Soviet Union is a predominantly low-lying country. At least half her territory is below the 200 m contour and barely one-tenth is more than 1000 m above sea level. As Map 3 indicates, lowlands predominate from the western border to the Yenisey valley, this zone comprising three great plains; the East European Plain (1)*, the West Siberian Lowland (2) and the Caspian-Turanian Lowland (3). These are separated only by the relatively weak relief barriers of the Ural Mountains (4) and the Kazakh Upland (5), and the western half of the country has been likened to a huge amphitheatre opening northwards to the Arctic and hemmed in by high mountains on its southern side.

The East European Plain (1) extends some 2000 km from the western border to the Urals and nearly 3000 km from the Arctic Ocean to the Caucasus. With the exception of the Khibiny Mountains in the Kola peninsula (6), which rise to just below 1200 m, the land rarely reaches 400 m, and is for the most part below 200 m. The most low-lying section lies to the north of the latitude of Moscow; this was a zone of glacial deposition and the low, hummocky surface is often poorly drained. In the southern section, conditions are a good deal drier and relief rather more pronounced; here there is a west-east alternation of low, dissected plateaus and broad valley plains comprising the Volyn Uplands (7), the Dnepr Lowlands (8), the Central Russian Elevation (9), the Don Lowlands (10), the pre-Volga Heights (11) and the Volga valley (12). Extensive plains below sea level to the north of the Caspian indicate the former extent of that water body and are linked to the plains of the southern Ukraine by the Azov-Caspian depression (13). To the north of the latter, the Donets Heights (14) rise fairly abruptly to about 350 m.

The East European Plain is bordered on its eastern side by the Ural Mountains (4). Although

* A figure in brackets refers to the key numbers on Map 3.

traditionally the boundary between Europe and Asia, the Urals are not a major barrier to movement between the two continents, consisting of a number of modest, broken ridges with numerous easy passes. In the north, where the ridges are close together and the whole system is barely 100 km wide, a number of peaks top 1,500 m. The central section is much lower and more broken and it is here that the main trans-Ural routes are to be found. In the south, the ridges fan out and the system is about 200 km across; the highest summit (1,640 m) occurs in this section, but crest lines are generally at 1,000 – 1,200 m.

East of the Urals lies the West Siberian Lowland (2), one of the most striking relief features of the USSR. In an area of well over 2.5 million sq. km, the land nowhere rises above 200 m and nearly half the plain is less than 100 m above sea level. Broad flood plains and vast, low-lying swampy lands are characteristic of this region save in the south, where drainage is better and the bulk of the settlement is to be found.

The third great lowland – the Caspian-Turanian Lowland (3) – is for the most part a basin of inland drainage towards the Aral Sea with desert or semi-desert landforms. Much of the plain is below the 200 m contour but there are a few low plateaus, such as the Ust-Yurt Plateau (15), which reach 300 m. In the south, the plain rises slowly to meet the mountain foothills at about 1,000 m.

The Caspian-Turanian and West Siberian lowlands are separated by the hill country of the Kazakh Upland (5), a plateau area 450 – 900 m above sea level with a few higher areas reaching 1,400 m.

The territory between the West Siberian and Lena lowlands constitutes the Central Siberian Plateau (16). Here, a series of erosion platforms have been cut indiscriminately across rocks of varied age to give a series of dissected plateaus at 350 – 700 m. In the north-west, the residual massif of the Putoran Mountains (17) reaches 1,700 m. On its eastern side, the Central Siberian Plateau is separated by the Lena basin (18) from the mountain ranges of the Far East.

The southern frontier zones of the European

USSR contain a number of young fold mountain areas. In the extreme west, Soviet territory now crosses the Carpathians (19) into the Hungarian basin, while the Crimean Mountains (20) front the Black Sea in the extreme south of the Ukraine. Both systems have crest lines between 1,200 and 1,600 m. The Caucasus (21) is a good deal more complex and comprises three main divisions. In the north, the Main Caucasian Range rises abruptly above the neighbouring plains to a crest line at 2,500 – 3,000 m across the Black Sea – Caspian isthmus. Numerous peaks exceed 4,000 m and the range includes the highest mountain in Europe – Mt. Elbruz – at 5,633 m. To the south, there is an equally abrupt descent to the Trans-Caucasian depression, a corridor between the two seas with deltaic lowlands at either end. South again are the largely volcanic Lesser Caucasus, rising in places to 4,000 m, which merge with the neighbouring mountain zones of Turkey and Iran.

In Soviet Central Asia, a series of high ridges – the Tyan Shan, Alay and others – run roughly east-west across the Sino-Soviet frontier, enclosing hill-foot and valley basins. Crest-lines above 3,000 m are common and the most southerly range, the Pamirs (22), includes the Soviet Union's highest peaks – Mt. Communism (7,500 m) and Lenin Peak (7,126 m).

Beyond the ancient routeway of the Dzungarian Gate (23), the southern mountain barrier continues in the Altay (24) and Sayan (25) ranges, the latter rising to nearly 3,000 m east of the upper Yenisey.

The lands around Lake Baykal, particularly in Trans-Baykalia (26), are characterised by dramatic mountain terrain resulting from block faulting on a massive scale. The summits of the uplifted blocks stand 1,800 – 2,000 m above sea level and are separated from the intervening rift valleys by steep fault scarps. One of these valleys contains Lake Baykal, the world's deepest lake, whose floor is 1,300 m below sea level – a drop of 3,300 m from the surrounding mountains.

The remainder of the Far East region is a land of complex fold mountain ranges and large river basins. Among the highest mountain ranges are the Stanovoy (27), which reach nearly 3,000 m, and those of the Kamchatka peninsula (28). In the latter there are several active volcanoes including Mt. Klyuchevsky (4,750 m). The most important lowlands are those of the Kolyma (29), fronting the Arctic, and the Amur-Ussuri (30) in the south, bordering Manchuria.

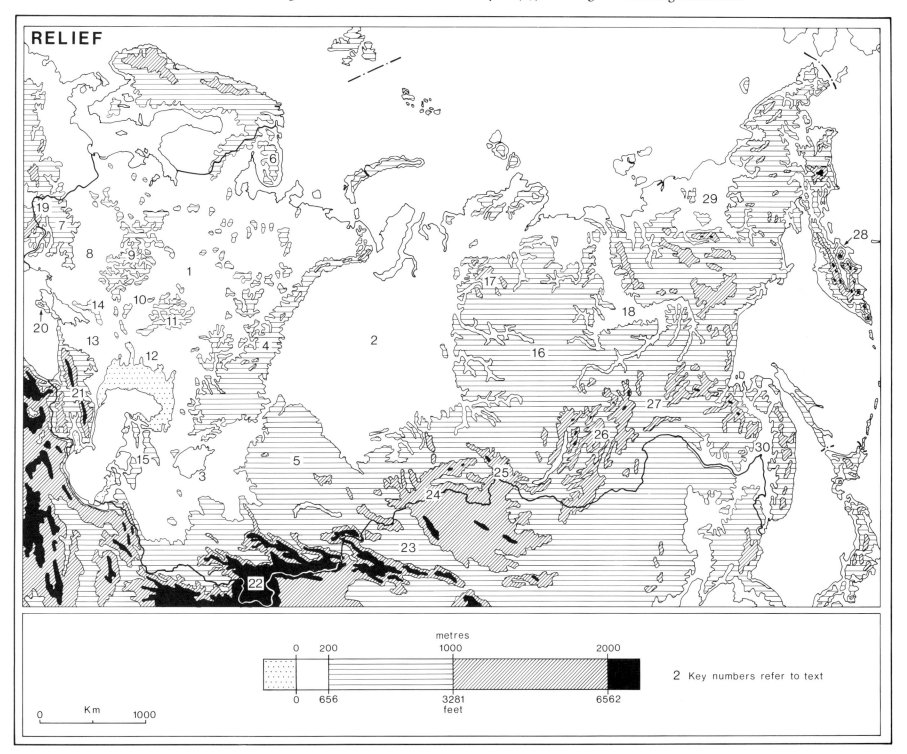

RELIEF

metres
0   200   1000   2000

2 Key numbers refer to text

0   656   3281   6562
feet

Km
0   1000

4 DRAINAGE

The main map shows the complete drainage pattern of the Soviet Union, while the inset map distinguishes the five major drainage systems – Pacific, Arctic, Inland, Baltic, Black Sea – and indicates the catchment areas of the major rivers.

A total area of nearly 13 million sq. km, about 58 per cent of the territory of the USSR, drains northwards to the Arctic Ocean, including the northern part of the European sector and virtually the whole of Siberia west of the watershed with the Pacific drainage. About two-thirds of this huge area is drained by the country's three biggest rivers – the Ob, Yenisey and Lena – whose catchment areas cover 3.2, 2.4 and 2.4 million sq. km respectively. All three rise in the southern mountain zone and follow courses of 3,000 km or more before debouching into the northern seas. 14 per cent of Soviet territory is drained by the biggest catchment of all, that of the Ob, which roughly corresponds to the West Siberian Lowland already described on page 6. This massive river, together with its major tributaries, the Irtysh and Tobol, has an extremely gentle gradient and flows slowly through its broad flood plain across the lowland. Like most Soviet rivers, the Ob is frozen in winter and the spring thaw comes first to its upper reaches in the south, often leading to massive flooding along the lower course.

Similar features characterise the main stream of the middle and lower Yenisey, but this river flows in a trough at the eastern edge of the West Siberian Lowland and thus has remarkably limited west bank tributary development. The major tributaries of the Yenisey are the Nizhnaya (Lower) and Podkamennaya (Stoney) Tunguska and the Angara, which drains Lake Baykal. These drain the western half of the Central Siberian Plateau, into which they are sharply incised.

The Lena catchment area shows considerable variety. In the south, it includes the block-faulted zone of Trans-Baykalia, where the alignment of the Lena itself and the upper reaches of the Vitim, Olekma and Aldan tributaries reflect the south-west to north-east grain of the country. The middle section of the Lena and its biggest tributary, the Vilyuy, flow through broad, open lowlands, while the lower course is somewhat confined between the Central Siberian Plateau, much of which drains into the Lena, and the fold mountains to the east.

A series of smaller, but still impressive, rivers, with extensive lowland basins form the remainder of the Arctic drainage. In the European north these are the Severnaya (Northern) Dvina (tribs. Sukhona, Vychegda) and Pechora and in north-east Siberia the Indigirka and Kolyma.

In marked contrast to the Arctic system, the Baltic drainage basin is the smallest of the five, though it covers nearly 600,000 sq km, and comprises, from west to east, the catchments of the Nyamunas (Niemen), Zapadnaya (Western) Dvina, Narva and Volkhov rivers.

The low watershed of the Valday Hills (343 m) and other minor upswellings on the surface of the East European plain separates the Baltic drainage from rivers flowing south to the Black Sea and Caspian. Some 1.4 million sq. km of Soviet territory drain into the Black Sea, including much of Belorussia, the whole of Moldavia and the Ukraine together with adjacent areas of the Russian republic and the western Caucasus. By far the most important rivers are the Dnepr and Don, each nearly 1,500 km in length. For the most part, these rivers flow in broad, open valleys, but the Dnepr is sharply incised across the Ukrainian massif in the Dnepr bend. Further west the Yuzhnaya (Southern) Bug and Dnestr rivers are incised into the Volyn uplands over much of their course. The western Caucasus has numerous small streams with steep gradients which flow swiftly to the Black Sea. Much of the drainage from the north face of the main range, however, is collected by the Kuban, which flows slowly to the Sea of Azov.

Despite the length of the Soviet Union's Pacific coastline, drainage to that ocean covers a limited area, since the main watershed runs close to the coast over much of its length, of about 2.2 million sq. km, some 10 per cent of Soviet territory. Thus much of the Pacific drainage consists of short rivers of a few hundred kilometres' length, often with steep gradients. There is, however, one major river system, that of the Amur in the south. Rising in Chinese Manchuria, the Amur forms the Sino-Soviet boundary for a length of some 2,000 km to the point where it is joined by the Ussuri, whence it flows north-eastwards for another 800 km to the sea. There are extensive lowlands in the Amur-Ussuri basin which, as already indicated, contain the bulk of the settled zone of the Soviet Far East.

Second only in size to the Arctic drainage system is the area of inland drainage (5.3 million sq. km) which is devided roughly equally between the Caspian and Central Asian systems. The former is dominated by the Volga, which, with a catchment of some 1.5 million sq. km, is second only to the three great Siberian rivers. The Volga (c. 2,500 km in length) with its east-bank tributary the Kama, drains a large part of the European plain and the west flank of the Urals southwards to the Caspian. Other important elements in the Caspian system are the Ural river and smaller rivers such as the Terek and Araks draining the eastern Caucasus.

The Central Asian system is a basin of inland drainage towards the Sea of Aral. Two major rivers – the Amu Darya and Syr Darya – rising in the southern mountain zone, reach the Aral sea, but lesser streams such as the Chu, peter out in the desert. As the map indicates, large sections of the Central Asian region are totally devoid of permanent watercourses.

The drainage patterns outlined above are significant in a variety of ways, notably for their effects on agriculture (Map 20), transport (Map 38) and the generation of hydro-electric power (Map 25).

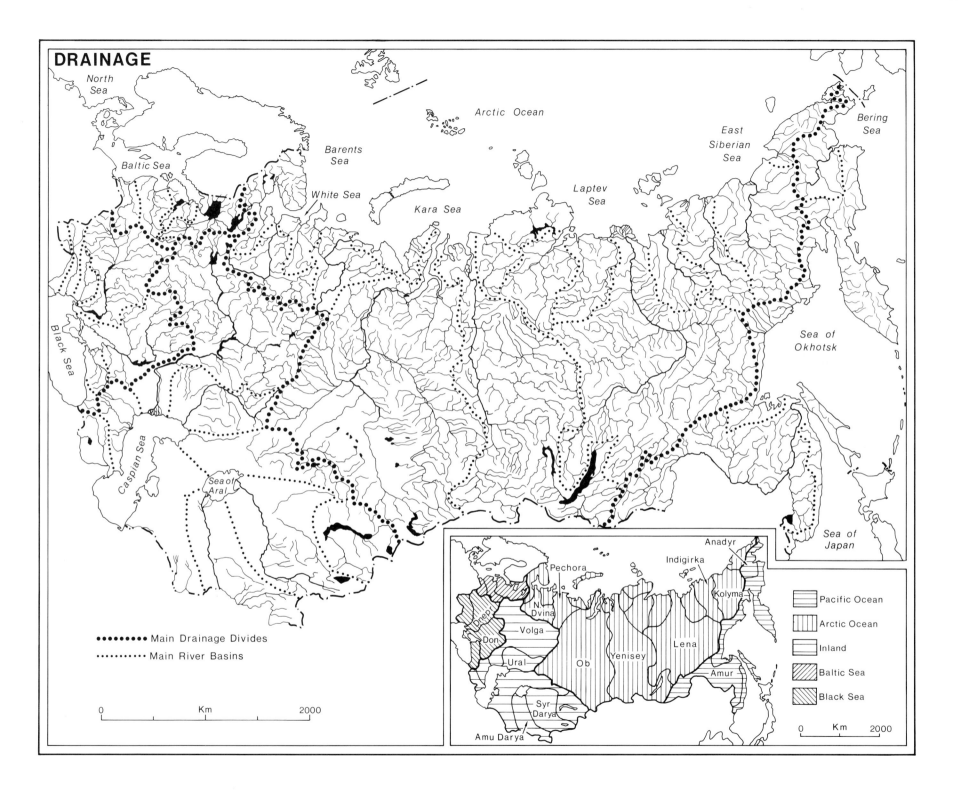

DRAINAGE

North Sea

Baltic Sea

Barents Sea

White Sea

Arctic Ocean

Kara Sea

Laptev Sea

East Siberian Sea

Bering Sea

Black Sea

Caspian Sea

Sea of Aral

Sea of Okhotsk

Sea of Japan

●●●●●●● Main Drainage Divides

······ Main River Basins

0   Km   2000

Anadyr

Indigirka

Pechora

Kolyma

N. Dvina

Dnep

Don

Volga

Lena

Ural

Ob

Yenisey

Amur

Syr-Darya

Amu Darya

Pacific Ocean

Arctic Ocean

Inland

Baltic Sea

Black Sea

0   Km   2000

9

5 and 6 CLIMATE

The maps and diagrams on pages 11 and 13 portray various aspects of the climates of the Soviet Union, which involve both strong regional contrasts and uniformity of climatic conditions over large areas.

Maps 5A and 5B Pressure and Winds

As a large, compact land mass, the USSR is dominated by continental climatic regimes. These display pronounced contrasts between the winter and summer seasons, with spring and autumn as brief periods of rapid transition between the two.

In winter (Map 5A), cooling of the Eurasian land mass results in the development of an intense high pressure cell over the interior of Asia with maximum January mean atmospheric pressure in excess of 1040 mb along the southern boundary of central Siberia. From this cell, a ridge of high pressure runs westwards towards central Europe; passing north of the Aral, Caspian and Black Seas, and a second ridge extends north-eastwards across Lake Baykal towards north-east Siberia and the Arctic. Pressure is lowest, though still above 1000 mb, in the extreme north-west and north-east of the country, which lie relatively close to the Icelandic and Aleutian low pressure systems respectively.

North of the western high pressure ridge, the prevailing winds blow from a south-western or southerly direction whereas, to the south of the ridge, they blow from the north or north-east. This arrangement has the effect of evening out the temperature variations which might be expected to occur with latitude. In the far eastern part of the country, cold winds blow outwards from the Siberian high pressure, giving low temperatures along the Pacific coast.

In summer (Map 5B), conditions are reversed. Low pressure develops over the continental interior and winds blow inwards, that is from the north-west or north over the western half of the country and from the east and south-east along the Pacific coast. The pressure gradient in summer is in any case a gentle one (there is a range of only 12 mb within Soviet territory as against 33 mb in winter) so that winds are often weak and their direction variable.

Maps 5C and 5D Temperatures

The pressure conditions and wind directions outlined above are reflected in the arrangement of the isotherms shown in Maps 5C and 5D, which are also influenced by the high-latitude situation of the country.

In winter (Map 5C), the great bulk of Soviet territory experiences January means below zero (°centigrade). The only exceptions are the extreme south of Central Asia and the Black Sea coasts of the Crimea and western Transcaucasia; in the latter cases, mountain barriers are a protection against cold winds from the north. Over most of the European USSR, latitude has relatively little influence on winter mean temperatures owing to the 'evening out' effect of the prevailing winds already mentioned, and isotherms run from north-west to south-east across the plains. Thus the January mean at Leningrad, for example ($-8°$C), is the same as that at Astrakhan in the Volga delta. The intensity of the winter cold increases towards the interior and parts of north-eastern Siberia experience January means below $-40°$C. The ameliorating effects of proximity to the sea are visible in the close packing of isotherms along the Pacific coast; however, because winds are out-blowing and because much of the coast is washed by cold currents from the north, this effect is limited and temperatures are abnormally low for a given latitude. Even in the extreme south of the Soviet Far East, which is in the same latitude ($43°$N) as the French Riviera, January means are around $-10°$C.

In summer (Map 5D), when pressure variations and wind direction are less important, isotherms reflect latitude and much of the complexity of the pattern is due to relief. Along the Arctic coast, July means are only 4 to $8°$C. There is a steady increase in temperatures towards the south with the highest means ($25-28°$C) in the Central Asian desert. There is, however, a marked diminution of temperatures along the Pacific coast owing to the influence of onshore winds.

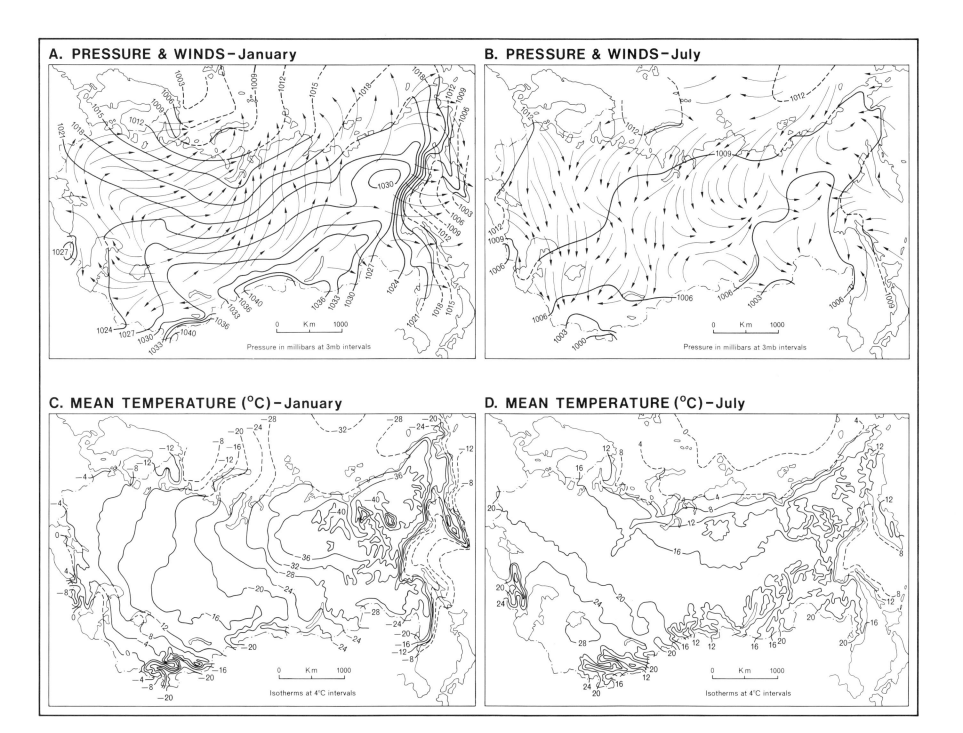

A. PRESSURE & WINDS–January

Pressure in millibars at 3mb intervals

B. PRESSURE & WINDS–July

Pressure in millibars at 3mb intervals

C. MEAN TEMPERATURE (°C)–January

Isotherms at 4°C intervals

D. MEAN TEMPERATURE (°C)–July

Isotherms at 4°C intervals

11

Map 6A Precipitation

Apart from the local effects of altitude, annual precipitation varies in response to the nature and movement of air masses and distance from the sea. These factors result in somewhat low precipitation totals over most of the USSR and a decline from the western boundary and the Pacific towards the interior.

Across the vast plains which constitute the western half of the country, annual totals decline from north-west to south-east. Totals above 600 mm are recorded in the Baltic republics, Belorussia and some western oblasts of the RSFSR, but the bulk of the European plain receives 450–550 mm and the West Siberian lowland 400–450 mm. Annual precipitation declines rapidly towards the interior of Soviet Central Asia and is only 50 or 60 mm in the zone to the south of the Aral Sea. The East Siberian plateau receives 300–400 mm and there is a further decline to about 250 mm in the Lena basin and even smaller amounts along parts of the Arctic coast. Annual totals rise again along the Pacific coast.

The various mountain areas of the USSR are, of course, characterised by considerably heavier precipitation than the adjoining lowlands, owing to the orographic effect, and there are a few areas where relief and wind direction combine to give really heavy rainfall. Most noteworthy in this respect is the small Black Sea coast zone of the Caucasus (humid western Transcaucasia), where the annual total exceeds 2000 mm at sea level and is even greater on seaward slopes as the result of the arrival of depressions from the west. In the Far East, onshore winds in summer result in quite heavy precipitation with annual totals of 600–700 mm at sea level and 1200 mm or more in the mountains.

Over most of the Soviet Union, precipitation shows a pronounced summer maximum, much of the rainfall occuring in convectional storms; exceptions are the Black Sea coastlands and much of Central Asia where a high proportion is derived from the eastward movement of winter depressions.

Despite the dominance of summer rainfall maxima, there is sufficient winter precipitation to give a considerable snowfall in most areas. Only a few Black Sea coastal lowlands are completely snow-free and even the Central Asian desert has a snow cover lasting from three to ten weeks. Further north, the duration of snow cover is, of course, much longer – $2\frac{1}{2}$–4 months over much of the European plain, 3–6 months in the West Siberian lowland and as much as 8 or 9 months along the Arctic coast.

Of particular significance to agriculture is the number of frost-free days, shown in Map 6B. Only in the extreme south of the country, in the southern Ukraine, the North Caucasus and Transcaucasian lowlands and in the southern part of Central Asia, does the frost-free period exceed six months (on average – in individual years it may be a good deal less), while over the interior of Siberia and the Arctic coast less than two months separate the last spring frost and the first frost of autumn.

The climatic diagrams printed with Map 6 show mean monthly temperature and precipitation for selected stations and give some idea of the range of climatic conditions within the USSR.

Sagastyr, near the mouth of the Lena river, represents conditions along the Arctic coast of Siberia. The mean temperature is above zero for only three months (June, July, August), with a maximum of only 5 °C in August and winter means below – 30°C from December to March. Far from any source of plentiful atmospheric moisture (the Arctic being ice-covered for long periods), Sagastyr has an annual precipitation of only 84 mm, three-quarters of which falls between June and September.

Leningrad, Turukhansk and Yakutsk indicate the increasing severity of the climate towards the interior. While Leningrad has seven months (April–October) with mean temperatures above freezing, this falls to five (May–September) at Yakutsk and only four (May–August) at Turukhansk, which is further north. Precipitation is relatively plentiful (475 mm) at Leningrad but only 442 mm at Turukhansk and 348 mm at Yakutsk. All three stations show a summer rainfall maximum; the wettest three months (July–September) have about 40 per cent of the annual total, while only 12–14 per cent falls in January–March.

Verkhoyansk is included as having one of the most extreme climates of any inhabited place on earth. The mean temperature is above zero for five months (May–September) with a maximum of 15°C in July. The January mean is – 50°C, giving an average range of 65°C between the hottest and coldest months. An absolute minimum of – 70°C (– 94°F or 126°F of frost) has been recorded at this station. The annual precipitation averages only 102 mm, two-thirds falling in June–August.

Moscow data indicate the less extreme but still severe conditions in the central part of the East European plain. There are seven months (April–October) with mean temperatures above freezing and the coldest month (January) registers – 10°C. Much lower daily minima are, of course, recorded (– 45°C in January 1979, for example). Precipitation is relatively plentiful at 534 mm and more evenly distributed, though the summer maximum is still visible. 37 per cent falls in June–August, 15 per cent in January–March. The latter is sufficient to give a deeper, though more short-lived snow cover than in Siberia.

Vladivostok represents the rather special conditions of the Pacific seaboard. The total precipitation (600 mm) is only 12 per cent above that of Moscow but the summer maximum is much stronger; 52 per cent falls in June–August. Winter temperatures are low for the latitude with a January mean of – 14°C.

Odessa illustrates conditions in the relatively favourable south-western part of the country. Only January and February have means below freezing (– 3°C, – 2°C or higher). The annual precipitation of 409 mm is fairly evenly distributed by Soviet standards.

Semipalatinsk lies in the steppes of eastern Kazakhstan and shows the much harsher conditions there than in the European steppe. Means are below freezing for five months (November–March), falling to – 16°C in January with a highest mean of 22°C in July. Annual precipitation is 185 mm.

Turtkul represents the southern part of the Central Asian desert. Monthly mean temperatures are below freezing in December—February, but the most striking feature is the summer heat with June—August means above 25°C. Annual precipitation is only 61 mm, three-quarters of which falls in January—April.

Yalta, Tbilisi, Baku and Batumi show the special conditions in the southern Crimea and Caucasus. These are among the few Soviet stations with no monthly mean temperature below zero and summers are hot with June—September means around 20°C. Annual rainfall at Yalta is 488 mm; there is a strong winter maximum with 76 mm in December. Tbilisi, in the Caucasian interior has a summer maximum but the winter maximum is again visible at Baku which, being on the drier side of the Caucasus, receives only 241 mm. Batumi, on the coast of western, humid Transcaucasia, is noteworthy for its extremely heavy rainfall, totalling 2367 mm. The driest month (May) has 71 mm and the wettest month (December) has 254 mm. The total for November—January is 823 mm, considerably more than any other lowland station in the Soviet Union receives in a whole year.

In summary, the USSR is dominated by cold continental climates with very low winter temperatures, considerable summer heat and a modest summer-maximum precipitation, exceptions occurring only in a few, relatively small peripheral areas.

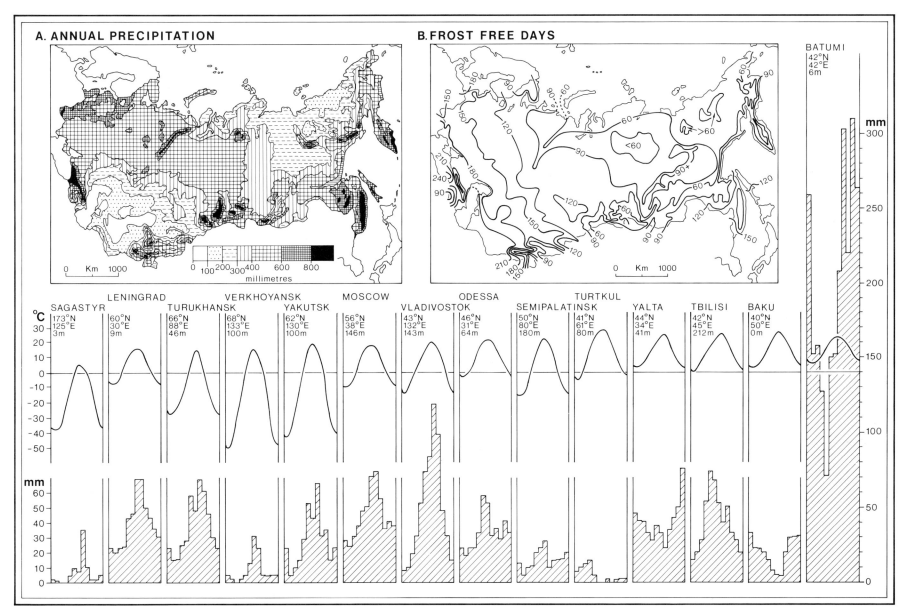

13

7 SOILS

Relief, drainage, climate, soils and natural vegetation (Maps 3–8) are closely inter-related elements of the physical environment which combine to produce the geographers' 'natural regions' (Map 9). Patterns of soil and natural vegetation are particularly closely related and there are strong similarities between Maps 7 and 8.

Map 7 shows the major soil types of the USSR. An initial division is that into 'lowland' and 'mountain' types. In the former, which cover the vast plains that dominate the western half of the USSR, relief affects soil type only at the level of localised variations and broad, latitudinal soil belts run from west to east across the country, reflecting changes in climatic conditions from north to south. In mountain areas, strong relief results in an altitudinal zoning of soil type. In such areas, which occupy the southern fringes of the western USSR and most of the territory to the east of Lake Baykal and the river Lena, considerable generalisation is necessary on a map at this scale.

In lowland areas, then, there is a north-south gradation of soil type from tundra and podzol in the north, through grey and brown forest soils, chernozems and chestnut soils to the desert soils of Central Asia; mountain areas have been simply classified according to the dominance of mountain tundra, mountain forest or mountain grassland soil types.

Tundra soils occupy a belt 100–500 km wide across the extreme north of the country from the Kola peninsula to the Bering Strait. Although rainfall amounts are small, the low evaporation rates prevailing over much of the year result in a moisture surplus so that tundra soils are poorly drained; the decay of organic matter is slow and the soils are highly acid. These features are intensified by the very subdued relief and by the presence of a permafrost layer a short distance below the surface. Tundra soils are in fact frozen for much of the year and only the top few metres thaw out in summer. A typical tundra soil has a shallow surface layer of raw, acid humus, beneath which there is a gley horizon resting on permafrost. Such soils are virtually useless for agriculture, which is in any case ruled out by the climate. Mountain tundra soils are widespread in northern Siberia, but in the south are confined to the higher summits.

Podzols are the most extensive soil type in the USSR, covering about half the national territory and roughly coinciding with the forest zone (see Map 8). The climate is such that, over the year as a whole, precipitation exceeds evaporation and the movement of soil moisture is predominantly downwards through the soil profile. Such soils are said to be 'leached'. The coniferous trees give rise to a highly acid raw humus layer at the surface. This decomposes slowly, producing soluble humic acids which percolate downards. Iron and calcium minerals are leached out of the upper horizon, which is therefore acid and pale in colour. Soluble materials are re-deposited at lower levels, where an iron-rich hardpan often develops, impeding drainage in the upper horizon. The strength of the leaching process varies with latitude, relief and parent material. Waterlogging often occurs, giving gleyed soils, while the excess of precipitation diminishes towards the south so that leaching becomes less intense. Organic content also increases southwards as the vegetation becomes more luxuriant. Thus, from north to south across the podzol zone, the soils become less acid, have a greater humus content and are more fertile. Most podzols in their natural state are poor in plant nutrient but can be successfully developed with modern farming techniques. Mountain forest soils, characteristic of most mountain areas of Siberia and the Far East, are similar, but steep slopes prevent the development of the mature podzol profile.

Grey and brown forest earths are a further stage in the southward progression. Leaching is much weaker than in the podzol zone and the organic content, developed from a predominantly deciduous forest cover, is much higher. These soils were important to the early development of Slav agricultural communities in the European plain (see p. 26).

Chernozem (black earth) soils are characteristic of the steppe grassland zone, running across the Ukraine, the Volga region and north Kazakhstan/south-west Siberia to the Altay and recurring in pockets further east. The black earth takes its name from its very dark upper horizon, often more than a metre thick, which is rich in humus derived from the thick grass cover. Winter frost and hot dry summers slow down the decomposition of organic matter and high evaporation rates prevent leaching so that humus accumulates. Calcium compounds are leached down in spring but are drawn upwards in summer and concentrated in a lime-rich horizon beneath the humus layer. Low acidity and high humus content combine to make chernozems extremely fertile.

Grassland soils (mountain meadow and steppe types) also predominate in southern mountain areas where summer temperatures are high and/or rainfall limited as in the Crimea, Caucasus and Central Asia, though in detail these areas have a great variety of soil types.

Chestnut soils occur to the south of the chernozem belt, reflecting drier, hotter conditions. Vegetation is poorer so that the organic content is less (hence the lighter colour); precipitation is less and evaporation greater, giving increased alkalinity. In places, salts accumulate to give highly alkaline *solonets* and *solonchak* soils, which are of little value.

Desert soils are the culmination of this progression. Vegetation is very poor so that the organic content is minimal and precipitation is very limited. Consequently, parent material has a major influence. Large areas of sand or clay desert have no true soil but, in the southern piedmont zone, *serozems* (grey earths) occur. Developed on loess, these have a moderate humus content derived from herbaceous vegetation and are very fertile when irrigated.

Red and yellow (sub-tropical) soils are a special type confined to the lowlands of western, humid Transcaucasia. These are clayey soils of considerable depth, produced by the weathering of mainly igneous rocks under the heavy year-round rainfall, hot summers and warm winters of this region. They are rich in iron and humus and highly productive.

SOILS

LOWLAND TYPES

Tundra soils

Podzols

Grey & brown forest earths

Chernozems

Chestnut soils

Desert soils

Red & yellow soils

MOUNTAIN TYPES

Mountain tundra soils

Mountain forest soils

Mountain meadow & steppe soils

Km
0   1000

15

8 VEGETATION

As already indicated, this map has strong similarities to the previous map (7) showing soils. Once again, broad latitudinal belts run right across the more low-lying parts of the country, contrasting with the more complex patterns of vegetation types in the mountainous south and east.

Tundra

Tundra vegetation, like tundra soils, occurs in a belt along the Arctic coast and in high mountain districts of Siberia. The vegetation cover is limited, and often discontinuous, comprising a variety of mosses, lichens, herbaceous plants, bushes and dwarf trees. The tundra is commonly divided into three sub-zones. The most northerly, the *Arctic tundra*, has much bare, unvegetated ground; mosses and lichens predominate and provide sustenance for the reindeer kept in this zone. Southwards, the *shrubby tundra* has mosses, lichens, herbaceous plants and dwarf trees such as the dwarf Arctic birch and the shrub willow; there are large areas of sphagnum bog. Finally, the *wooded tundra* comprises an alternation of tundra species with stunted birch, larch and spruce and is transitional to the forest zone.

Tayga

Forests of various types cover roughly half the territory of the USSR and by far the largest forest zone is that of the tayga. This is commonly divided into two parts along the line of the Yenisey river. The *western tayga*, in the East European plain and West Siberian lowland, is the more varied. Here, the main coniferous species are spruce and fir on the moister, heavier soils, forming a dense forest with poor undergrowth, and pine with a variety of shrubs and grasses on lighter, more heavily leached podzols. The *eastern tayga* is dominated by the larch, a species more tolerant of the extreme cold of Siberia, which has a shallow root system which allows it to grow above the permafrost layer.

The tayga is by no means a continuous area of coniferous forest. Large stands of birch, alder and willow occur and there are very large areas of peat bog developed from sphagnum and cotton sedge, especially in the northern part of the East European plain and in the West Siberian lowland. In drier areas, such as the middle Lena basin, there are stretches of coarse grassland.

Agriculture has penetrated the tayga only to a limited extent (see page 44) and its main economic function is as a source of timber. Large areas have been felled in the European section; in Siberia, lumbering occurs mainly in the south, but is now extending northwards.

Mixed and Deciduous Forest

South of the tayga is a zone in which coniferous and deciduous species both occur, the latter becoming more numerous and eventually dominant with decreasing latitude. The mixed and deciduous forest zone is triangular in shape, narrowing towards the Urals. Oak and spruce are the dominant tree species, but a variety of other types occur including ash, aspen, birch, elm, hornbeam, maple and pine. East of the Urals, a belt of birch/aspen woodland separates the tayga from the wooded steppe. Much of the mixed and deciduous forest zone has been cleared for agriculture, particularly in the west.

Far Eastern deciduous forest is the natural vegetation of the Amur-Ussuri lowlands where summer monsoonal rainfall supports luxuriant growth of oak, maple, ash and other species – an extension of the Manchurian forest zone.

Steppe

As the climate becomes drier and summers hotter, southwards across the European plain, forest gives way to steppe. The *wooded steppe*, as the name implies, is transitional between the two. Stands of oak and other deciduous species in the European section and of birch and aspen in the West Siberian lowland are interspersed with areas of grassland, which become larger towards the south.

Steppe grassland extends in a continuous belt across the western half of the country from Moldavia and the Ukraine to south-west Siberia and northern Kazakhstan, reaching the foothills of the Caucasus between the Black and Caspian seas. In southern Siberia there are pockets of wooded steppe and steppe in lowland basins. Trees are virtually absent, save along water-courses, and the vegetation is various types of grassland. From north to south the character of the grassland changes, becoming less luxuriant with an increasing proportion of xerophytic species.

Much of the steppeland has come under the plough and little natural grass cover remains to the west of the Volga. In the eastern steppe, though cultivation has expanded greatly during the past 100 years, sizeable areas of natural grassland remain.

Desert

The steppe grades southwards through the dry steppe semi-desert to true desert where, over large areas, there is little or no vegetation. Where vegetation does occur, the cover is usually discontinuous and involves xerophytic grasses, shrubs and saxaul. Grasses and sedges are more plentiful in the loess-covered piedmont zone.

Sub-tropical Forest

This is a specialised vegetation type developed under the special climate of humid western Transcaucasia. Dense woodland, much of which has been cleared for agriculture, is dominant, with oak, hornbeam, beech, holly, box and various climbing plants such as ivy and lianas.

Mountain Vegetation

In the areas classified as 'mountain types' in Map 8, there is a marked altitudinal zoning of natural vegetation. Coniferous forest is dominant in mountain areas of Siberia and the Far East, with tundra vegetation at high altitudes, while in the much drier mountain areas of Central Asia mountain grassland predominates. The western mountain areas of the Crimea and Caucasus have their own special, mainly broad-leaved, forest with much oak and beech at lower levels, above which are coniferous mountain forests, giving way eventually to high altitude alpine grassland. On the higher ranges of the Caucasus and the Central Asian mountains there is much bare rock and perennial snow.

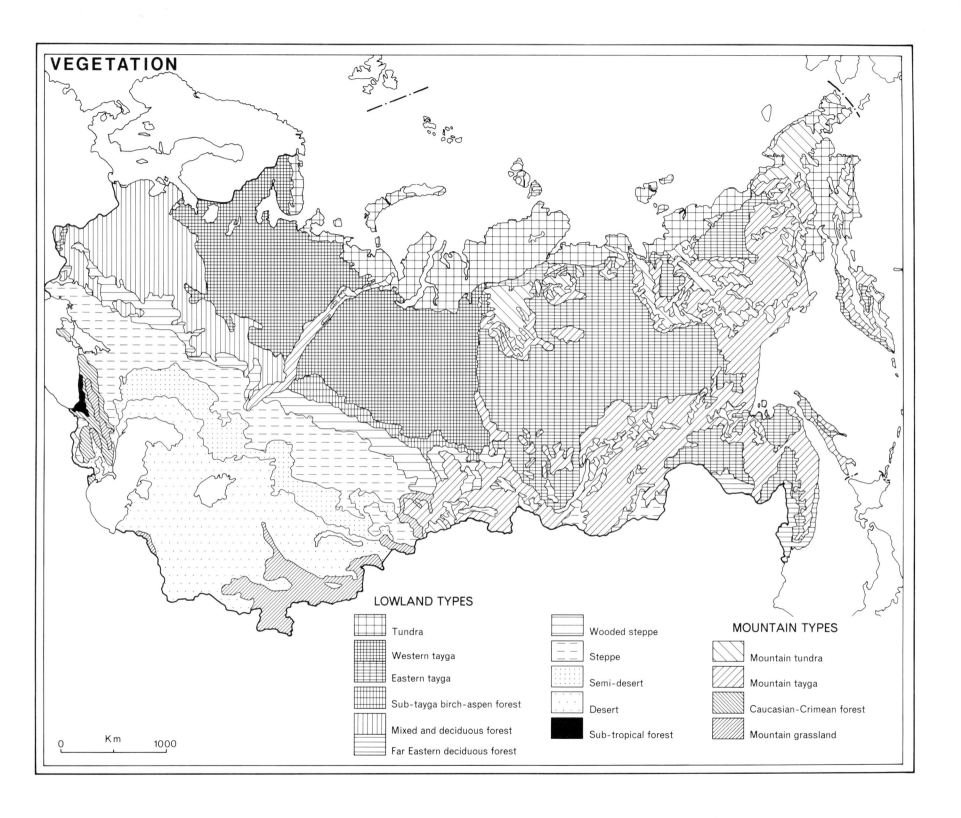

VEGETATION

LOWLAND TYPES

- Tundra
- Western tayga
- Eastern tayga
- Sub-tayga birch-aspen forest
- Mixed and deciduous forest
- Far Eastern deciduous forest
- Wooded steppe
- Steppe
- Semi-desert
- Desert
- Sub-tropical forest

MOUNTAIN TYPES

- Mountain tundra
- Mountain tayga
- Caucasian-Crimean forest
- Mountain grassland

Km
0   1000

9 NATURAL REGIONS

Soviet geographers have devoted much effort to the division of their country into natural regions as a means of illustrating the great variety of physical environments within the boundaries of the USSR. The scheme reproduced in Map 9 is that used in the great *Fiziko-Geograficheskiy Atlas Mira* (Moscow, 1964).

The divisions shown on this map are based on structure and relief combined with natural vegetation. The primary division is into 'Natural Countries' (*prirodnye strany*), here labelled 'Physiographic Regions' and is based on the structure and relief elements already discussed on pages 4 and 6, including such units as the East European plain, West Siberian lowland, Central Siberian plateau, Baykalia and the Yakut basin.

A second level of regionalisation is based on 'natural vegetation complexes' already discussed on page 16. 'Plains Types' (*ravinnye territorii*) show a dominance of latitudinal zonation: the East European plain, for example, being divided from north to south into areas characterised by tundra, wooded tundra, northern tayga, central tayga, southern tayga and mixed forest, deciduous forest and wooded steppe, and steppe vegetation complexes. 'Mountain Types' (*gornye territorii*) have a dominance of altitudinal zoning and may cover the whole or part of a 'natural country'. Thus the Amur-Maritime country (18) contains areas of plain with central tayga, southern tayga/mixed forest and deciduous forest/wooded steppe vegetation, and mountain areas dominated by mountain forest and meadow.

The full scheme uses these vegetation-based divisions plus a third level of regionalisation based on local relief, drainage and vegetation to distinguish a set of 'natural provinces' (*prirodnye provintsii*). Numbering 194 in all, these natural provinces are too small and too numerous to be shown clearly and labelled on Map 9, but their boundaries are shown on the small Map 9A below. Lack of space precludes a full listing of

these units, but the following examples will give some idea of the nature of the scheme.

The 'natural country' of the East European plain has a large central zone distinguished by 'southern tayga and mixed forest' vegetation. This is divided into six 'natural provinces'. In the north-west is the 'Baltic moraine-hill lake plain with mixed forest' which gives way southwards to the 'swampy outwash plain of the Polesye with pine/broadleaf forest'. The north-centre, from Smolensk and Moscow to just south of Lake Onega is the 'Valday-Smolensk-Moscow upland with moraine hills and mixed forest', south of which lies the 'northern Central Russian Elevation with broadleaf and coniferous sub-tayga forest'. East of these two central 'provinces' is the 'upper Volga plain with swampy lowland in the mixed forest zone' and east again the 'Vyatka-Kama south-tayga upland plain'.

The Ural 'natural country', which includes Novaya Zemlya, contains, from north to south, the following ten 'natural provinces'; 'middle elevation mountains of Novaya Zemlya's north island with extensive ice sheets and arctic tundra'; 'Arctic tundra low mountains and hilly plains of southern Novaya Zemlya'; 'Pay-Khoy residual hilly tundra upland'; 'Polar Urals: middle-elevation mountains with alpine relief forms and denuded-summit tundra vegetation'; 'sub-polar Urals with Alpine relief forms and north-tayga forest at its foot'; 'middle-elevation Northern Urals with middle-tayga forest at foot'; 'low mountain and hill-ridge south-tayga Middle Urals'; 'wooded steppe South Urals with mountain-tayga slopes and alpine meadows on summits'; 'trans-Ural steppe peneplanised plain'; and 'Mugodzhar desert-steppe residual upland'.

Such a detailed scheme of physical regionalisation provides a vast amount of information on the physical environment of the USSR. So large is the Soviet Union, however, that even the 'small' natural provinces have an average area of no less than 115,475 sq. km, roughly equivalent to that of England.

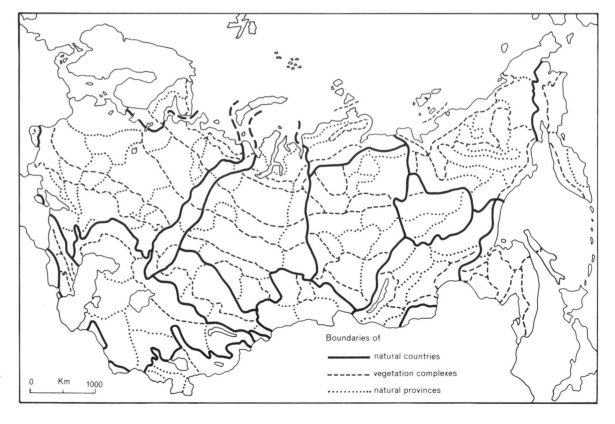

Boundaries of

—————— natural countries

- - - - - - vegetation complexes

·············· natural provinces

0   Km   1000

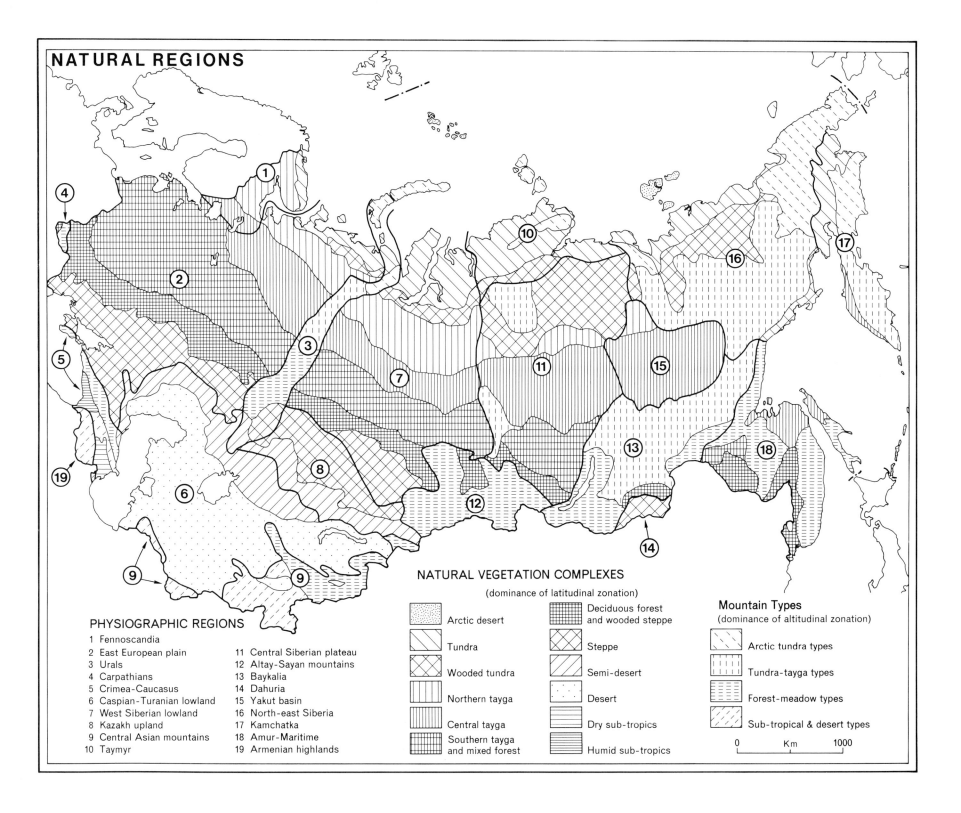

NATURAL REGIONS

PHYSIOGRAPHIC REGIONS

1 Fennoscandia
2 East European plain
3 Urals
4 Carpathians
5 Crimea-Caucasus
6 Caspian-Turanian lowland
7 West Siberian lowland
8 Kazakh upland
9 Central Asian mountains
10 Taymyr
11 Central Siberian plateau
12 Altay-Sayan mountains
13 Baykalia
14 Dahuria
15 Yakut basin
16 North-east Siberia
17 Kamchatka
18 Amur-Maritime
19 Armenian highlands

NATURAL VEGETATION COMPLEXES
(dominance of latitudinal zonation)

- Arctic desert
- Tundra
- Wooded tundra
- Northern tayga
- Central tayga
- Southern tayga and mixed forest
- Deciduous forest and wooded steppe
- Steppe
- Semi-desert
- Desert
- Dry sub-tropics
- Humid sub-tropics

Mountain Types
(dominance of altitudinal zonation)

- Arctic tundra types
- Tundra-tayga types
- Forest-meadow types
- Sub-tropical & desert types

0   Km   1000

PART 2 HUMAN GEOGRAPHY

10 POLITICAL DIVISIONS

As indicated in our discussion of the ethnic composition of the Soviet population (Map 16, p. 35), the USSR is a multinational state embracing a great diversity of ethnic groups. This situation is recognised in the Soviet constitution, in the organisation of the country's government and in the delineation of territorial sub-divisions.

The political-administrative divisions shown in Map 10 are based first and foremost on the distribution of the various ethnic groups. Each of the units shown is considered to be the national homeland of the ethnic group after which it is named – the 'titular' group –, though no ethnic group is in any way confined to its homeland, and no political-administrative division is inhabited solely by members of the titular group. Long-continued migration movements, both before and since the revolution, have ensured that all areas have a population which is to some extent ethnically diverse. In many cases, particularly among the lower grade political divisions, the titular group is outnumbered by members of other nationalities.

These political divisions are directly represented in the Soviet legislature, each sending its own delegates to the Soviet of Nationalities, one of the two 'houses' which constitute the Supreme Soviet of the USSR, the country's top legislative body. (The other house is the Soviet of the Union, where delegates are elected on a constituency basis, roughly one per 300,000 people.)

The system of political-administrative divisions is a hierarchical one. At the top are the Union Republics – the Russian Soviet Federated Republic and the other 14 Soviet Socialist Republics (SSRs). In theory, the USSR is a voluntary federation among these 15 republics. As the inset diagram shows, SSRs vary enormously in population size, ranging from the RSFSR with 137.5 million inhabitants to the Estonian SSR with only 1.5 million. Regardless of population size, the SSRs are equally represented in the Soviet of Nationalities, to which each sends 32 delegates. Each Union Republic has its own Supreme Soviet responsible for internal affairs and each, in theory, has the right to secede from the Soviet Union. This affects the geographical pattern of SSRs; each must have a boundary with the outside world to prevent the possibility of the 'political anomaly' of an independent state completely surrounded by Soviet territory which might otherwise result from the secession of an SSR situated in the interior of the country. Thus, although Union Republics are said to be the homelands of the larger and more advanced ethnic groups, certain large groups, like the Tatars, whose homelands are in the interior, can never achieve SSR status.

Subordinate to the Union Republics in which they lie are the Autonomous Soviet Socialist Republics (ASSRs), at present 20 in number. The distribution of these units reflects the presence of important minority groups, clustered particularly in the Caucasus and the zone between the middle Volga and the Urals. Again there is a wide population range – the Tatar ASSR (26) has 3.4 million inhabitants, the Tuvinian ASSR (27) only 266,000 – but each sends 11 delegates to the Soviet of Nationalities.

Finally, Autonomous Oblasts and Autonomous Okrugs are the home areas of relatively small minority groups. Autonomous Oblasts are administrative areas on a par with the standard oblast (see Map 11) but also send five delegates each to the Soviet of Nationalities. Autonomous Okrugs are subordinate to the administrative oblast or kray, but send one delegate. The latter are confined to the RSFSR, occurring mainly in northern Siberia, where they form large areas of thinly settled territory, the homelands of small indigenous ethnic groups like the Koryaks (48) and Evenki (45).

Following is a list of the 53 political-administrative areas shown on Map 10

Soviet Socialist Republics (SSRs): 1. Russian Soviet Federated Socialist Republic (RSFSR); 2. UK. Ukrainian SSR; 3. BE. Belorussian SSR; 4. UZ. Uzbek SSR; 5. Kazakh SSR; 6. GE. Georgian SSR; 7. AZ. Azerbaydzhanian SSR; 8. LI. Lithuanian SSR; 9. MO. Moldavian SSR; 10. LA. Latvian SSR; 11. KI. Kirgiz SSR; 12. TA. Tadzhik SSR; 13. AR. Armenian SSR; 14. TU. Turkmen SSR; 15. ES. Estonian SSR;

Autonomous Soviet Socialist Republics (ASSRs): 16. Bashkir ASSR; 17. Buryat ASSR 18. Dagestan ASSR; 19. Kabardino-Balkar ASSR; 20. Kalmyk ASSR; 21. Karelian ASSR; 22. Komi ASSR; 23. Mariy ASSR; 24. Mordov ASSR; 25. North Osetin ASSR; 26. Tatar ASSR; 27. Tuvinian ASSR; 28. Udmurt ASSR, 29. Chechen-Ingush ASSR; 30. Chuvash ASSR; 31. Yakut ASSR; 32. Kara-Kalpak ASSR; 33. Abkhaz ASSR; 34. Adzhar ASSR; 35. Nakhichevan ASSR;

Autonomous Oblasts (A. Obs): 36. Gorno-Altay A.Ob.; 37. Adygey A.Ob.; 38. Khakass A.Ob.; 39. Karachayevo-Cherkess A.Ob.; 40. Yevreysk (Jewish) A.Ob.; 41. South Osetin A.Ob.; 42. Nagorno-Karabakh A.Ob. 43. Gorno-Badakhshan A.Ob.;

Autonomous Okrugs (A.Oks): 44. Taymyr (Dolgano-Nenets) A.Ok.; 45. Evenki A.Ok.; 46. Nenets A.Ok.; 47. Ust-Orda Buryat A.Ok.; 48. Koryak A.Ok.; 49. Chukot A.Ok.; 50. Komi-Permyak A.Ok.; 51. Yamalo-Nenets A.Ok.; 52. Khanty-Mansiy A.Ok.; 53. Aga-Buryat A.Ok.

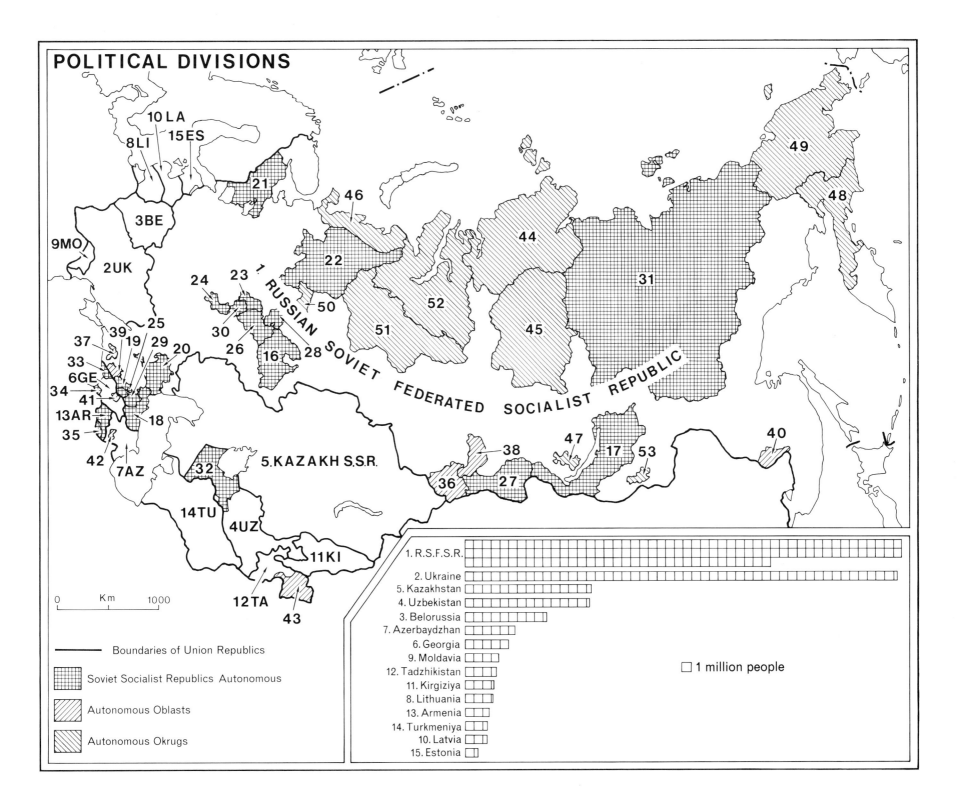

POLITICAL DIVISIONS

10 LA
8 LI
15 ES
21
3 BE
9 MO
2 UK
46
22
44
49
48
24
23
1. RUSSIAN SOVIET FEDERATED SOCIALIST REPUBLIC
50
52
31
51
45
25
39
19 29 20
37
30
26 16 28
33
6 GE
34
41
18
13 AR
35
47
17 53
42
7 AZ
38
40
36 27
32
5. KAZAKH S.S.R.
14 TU
4 UZ
11 KI
12 TA
43

0   Km   1000

— Boundaries of Union Republics

Soviet Socialist Republics Autonomous

Autonomous Oblasts

Autonomous Okrugs

1. R.S.F.S.R.
2. Ukraine
5. Kazakhstan
4. Uzbekistan
3. Belorussia
7. Azerbaydzhan
6. Georgia
9. Moldavia
12. Tadzhikistan
11. Kirgiziya
8. Lithuania
13. Armenia
14. Turkmeniya
10. Latvia
15. Estonia

☐ 1 million people

11 ECONOMIC AND ADMINISTRATIVE DIVISIONS

In addition to its organisation into the 53 political units portrayed in Map 10, the USSR is also subdivided into economic regions and administrative areas. The various systems of regionalisation are closely linked. Each economic region is an aggregate of several administrative and/or political units, and the boundaries of economic regions never cut across those of the political divisions*. This has led to a number of anomalies, the most striking of which is the division of the Donbass coalfield between the Donets-Dnepr region of the Ukraine and the North Caucasus region of the RSFSR.

There are two levels of economic and administrative regionalisation. The lower-level units are known as *Economic-Administrative Areas* and at present number 155. The majority of Union Republics are divided into *oblasts*, which are the standard administrative units; each consists of an organising urban centre (shown by a dot on Map 11) from which, in most cases, the oblast takes its name, and the territory attached to it (e.g. Moscow oblast, Kiev oblast, etc.). The RSFSR also contains a number of *krays*, which perform similar administrative functions; the distinction between oblast and kray has become blurred over the years. *Autonomous oblasts* and *Autonomous okrugs*, which are ethnically-based political units, are administratively subordinate to the kray or oblast in which they occur.

Autonomous Soviet Socialist Republics (ASSRs), as well as being directly represented in the Soviet of Nationalities, have the same local

* There is one special case. The Kaliningrad oblast, annexed from Germany in 1945, is a detached part of the RSFSR which has been attached to the three Baltic republics to form the Baltic economic region.

government functions as the oblasts and krays. In addition, a number of smaller SSRs which are not subdivided (Estonia, Latvia, Lithuania, Moldavia, Armenia) are economic-administrative areas as well as being political units.

These lower-tier units – oblasts, krays, ASSRs and small SSR's – are grouped to form the larger units known as *Major Economic Regions*. The boundaries of these have been altered on several occasions but the present pattern has remained virtually unchanged since the early 1960s. There are now 19 Major Economic Regions. The RSFSR is divided into ten – North-West, Centre, Volga-Vyatka, Black Earth Centre, Volga, North Caucasus, Ural, West Siberia, East Siberia, Far East – and the Ukraine into three – Donets-Dnepr, South-West, South. Three Major Economic Regions consist of groups of SSRs – the Baltic region comprising Estonia, Latvia, Lithuania and the Kaliningrad oblast of the RSFSR, the Transcaucasus region consisting of Georgia, Armenia and Azerbaydzhan, and the Central Asian region which unites the Uzbek, Kirgiz, Tadzhik and Turkmen republics. Kazakhstan and Belorussia are Major Economic Regions in their own right, while Moldavia, officially described as 'a republic outside the system of major economic regions', is the nineteenth unit.

The Major Economic Regions vary greatly in size and population and, like the administrative divisions, tend to be large in territory and relatively small in population in the thinly settled parts of the country, while the reverse is true in the more densely populated areas. This feature is illustrated by the inset cartogram, where the size of each region is based on its population rather than its area. The large populations of the central European regions and the small populations of the vast Siberian territories are striking features.

In the late 1950s, considerable economic powers were devolved to the Major Economic Regions and Economic-Administrative Areas, but the 1960s and 1970s have seen a return to centralised economic control. The Economic-

Administrative Areas are now essentially units of local government only. The Major Economic Regions are used for planning purposes by the planning organisations of the USSR and the constituent republics and as 'standard regions' for the publication of a variety of statistics; Soviet geographers frequently use these divisions as the basis for regional description and analysis.

Following is a list of the Major Economic Regions shown on Map 11 and their relationship to the 15 Union Republics.

USSR (**22,402,200 sq. km; pop. 262,442,000**)

RUSSIAN SFSR (17,075,400 sq. km; pop. 137,552,000)

North-Western Economic Region (1,662,800 sq. km; pop. 13,275,000)
Central Economic Region (485,100 sq. km; pop. 28,947,000)
Volga-Vyatka Economic Region (263,300 sq. km; pop. 8,343,000)
Central Black Earth Economic Region (167,700 sq. km; pop. 7,797,000)
Volga Economic Region (680,000 sq. km; pop. 19,393,000)
North Caucasus Economic Region (355,100 sq. km; pop. 15,487,000)
Ural Economic Region (680,400 sq. km; 15,568,000)
West Siberian Economic Region (2,427,200 sq. km; pop. 12,959,000)
East Siberian Economic Region (4,122,800 sq. km; pop. 8,158,000)
Far Eastern Economic Region (6,215,900 sq. km; pop. 6,819,000)
UKRAINIAN SSR (603,700 sq. km; pop. 49,757,000)
Donets-Dnepr Economic Region (220,900 sq. km; pop. 21,045,000)
South-Western Economic Region (269,400 sq. km; pop. 21,578,000)
Southern Economic Region (113,400 sq. km; pop. 7,134,000)
Baltic Economic Region (189,100 sq. km; pop. 8,192,000)
ESTONIAN, LATVIAN and LITHUANIAN SSRs and Kaliningrad oblast of the RSFSR
Transcaucasian Economic Region (186,100 sq. km; pop. 14,075,000)
ARMENIAN, AZERBAYDZHANIAN and GEORGIAN SSRs
Central Asian Economic Region (1,277,100 sq. km; pop. 25,480,000)
KIRGIZ, TADZHIK, TURKMEN and UZBEK SSRs
Kazakhstan Economic Region (2,717,300 sq. km; pop. 14,685,000)
KAZAKH SSR
Belorussian Economic Region (207,600 sq. km; pop. 9,559,000)
BELORUSSIAN SSR
MOLDAVIAN SSR (33,700 sq. km; pop. 3,948,000)

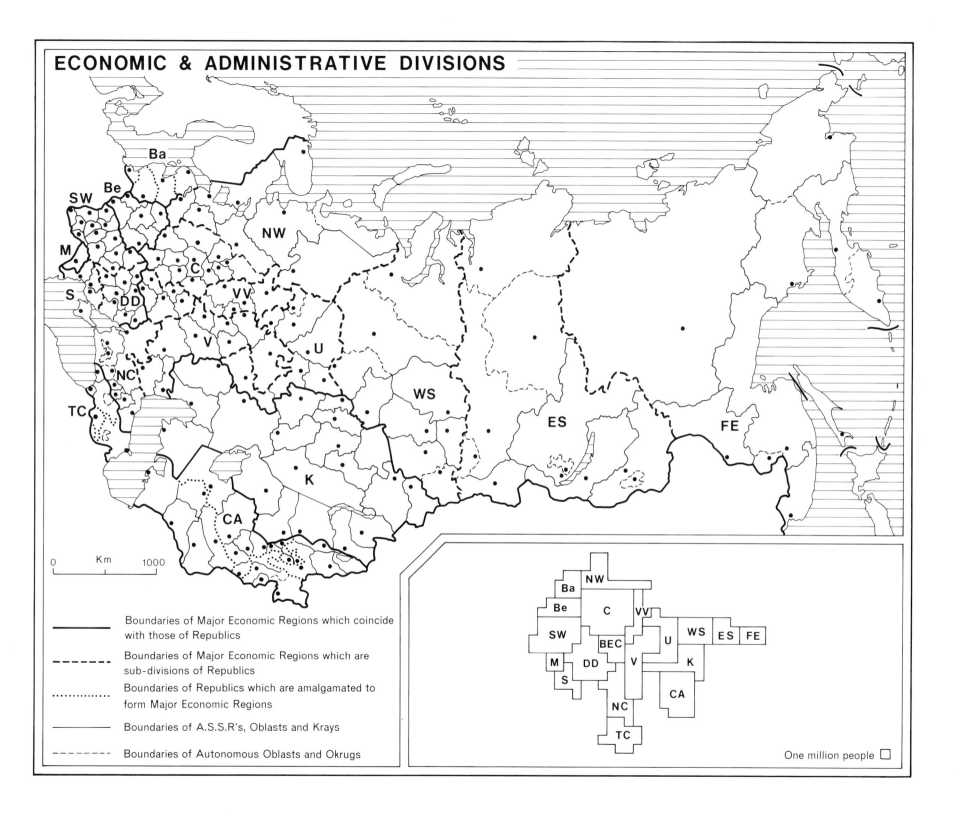

ECONOMIC & ADMINISTRATIVE DIVISIONS

0   Km   1000

Boundaries of Major Economic Regions which coincide with those of Republics

Boundaries of Major Economic Regions which are sub-divisions of Republics

Boundaries of Republics which are amalgamated to form Major Economic Regions

Boundaries of A.S.S.R's, Oblasts and Krays

Boundaries of Autonomous Oblasts and Okrugs

One million people □

12 GROWTH OF THE STATE

The present territory of the Soviet Union was for the most part inherited in 1917 from its political predecessor, the Russian Empire. Like the other imperial domains of the European powers, the Russian Empire was built up by more than 300 years of continuous expansion from the sixteenth century to the First World War. There are, however, two major contrasts with the empires of the western powers. While the latter were built up by overseas exploration and territorial acquisition and consisted of blocks of territory scattered across the globe, the Russian Empire grew by overland expansion and developed as a single territorial unit. Furthermore, whereas the west European empires have largely disappeared as their constituent parts have achieved independence, the Russian Empire has remained virtually intact in the guise of a Union of Soviet Socialist Republics.

Prior to the sixteenth century, 'Russia' consisted of a series of separate Slav principalities occupying the forested zone of the east European plain. In the middle of the tenth century, a loose confederation, centred on Kiev and known as 'Kievan Rus', was established and lasted some 300 years. In the mid-thirteenth century, however, there occurred the last and most devastating nomadic incursion from the east, that of the Tatars, and the Slav lands became part of a vast Tatar-Mongol empire, stretching from eastern Europe to the shores of the Pacific.

Although the Tatars took physical control of the steppelands, the Slav principalities in the forest zone were, after the initial attack, left relatively undisturbed, subject only to the payment of tribute to their Tatar overlords. The Russian principalities grew in wealth and power and were eventually able to regain their independence. The first to do so was that centred on Moscow (Muscovy). In 1476, Ivan III, ruler of Muscovy, ceased to pay tribute to the Khans and defeated a Tatar army sent to punish him. At the same time, Muscovy was extending its territory at the expense of the other Slav principalities, and in 1547 Ivan IV was declared 'Tsar of all the Russias', setting the seal on the supremacy of Moscow. From then onwards, the growth of the Russian Empire may be viewed as an outward expansion from the Moscow core.

In the sixteenth and seventeenth centuries, this expansion was mainly towards the east. By 1556, Russian territory extended south along the Volga to the Caspian and by 1600 it reached across the Urals into western Siberia. Progress across Siberia during the seventeenth century was particularly rapid. Russian explorers reached the Pacific in 1639 and by 1690 controlled its shores from the Bering Strait to the north of the Amur. In the far south-east, however, Russian expansion was checked by the growth of China, and the Treaty of Nerchinsk (1689) fixed the Russo-Chinese border along the crest of the Stanovoy range. In the middle of the nineteenth century, when China was weak, Russian territory was extended southward to the present boundary along the Amur and Ussuri, established in 1859. Thereafter, further Russian expansion in this sector was halted by the growth of Japan.

In the far north-east, however, there was no check to the growth of the Russian Empire, which crossed the Bering Strait and extended down the west coast of North America to a point near present-day San Francisco. These North American territories were abandoned in 1867, when they were sold to the United States for $7.2 million.

The expansion described so far was mainly in the forest zone of Siberia, where the environment was familiar to Russian settlers and the indigenous population small and poorly organised. Movement southward into Central Asia, where the Moslem Khanates were protected by a vast desert barrier, was more difficult. Expansion in this direction was due in part to fears of British expansion north from India, and Central Asia came under Russian control mainly during the nineteenth century; a southern frontier, with Afghanistan as a buffer state, was agreed in 1888.

The Caucasus was another region acquired in the nineteenth century. The northern shores of the Black Sea were annexed by 1800, but it was not until 1878 that the Russians subdued the numerous fiercely independent peoples of the Caucasus and the frontier with Turkey and Iran was fixed in roughly its present position.

The European west was a zone of particular difficulty and danger, for here Russia was confronted by a succession of hostile powers anxious to prevent her expansion towards western Europe. The lands around the Baltic were annexed in the early eighteenth century following Peter the Great's victory over Sweden, and large territories were gained from the partitions of Poland in 1772 and 1793; the Duchy of Warsaw was annexed at the end of the Napoleonic Wars in 1815.

Following the First World War, the newly established USSR lost large territories in the west. Finland and the Baltic States became independent, and Poland was re-established. During and after the Second World War, however, the Baltic States were regained, there was westward expansion at the expense of Poland and the USSR gained small but strategically important territories never previously under Russian control in former East Prussia and the Sub-Carpathian Ukraine (Ruthenia), annexed from Czechoslovakia.

Thus the present-day territory of the Soviet Union is roughly equal to that of the Russian Empire at its maximum extent. Only Alaska, Finland and certain Polish lands have been lost.

GROWTH OF THE STATE

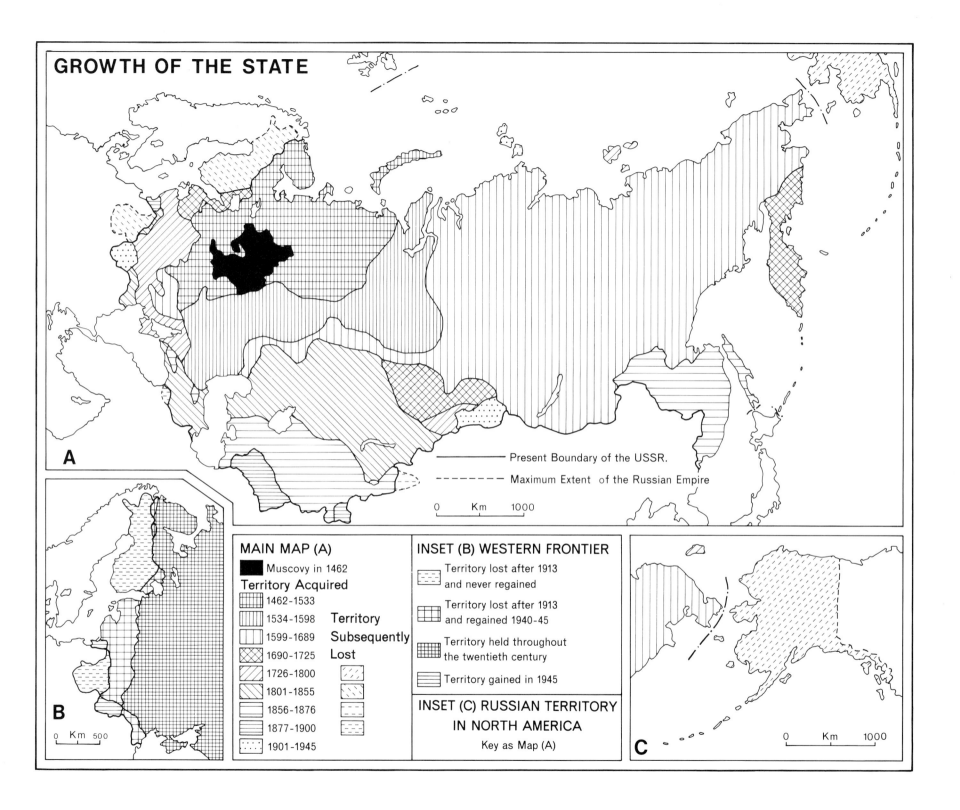

A

——————— Present Boundary of the USSR.

— — — — Maximum Extent of the Russian Empire

0   Km   1000

B

0   Km   500

MAIN MAP (A)

■ Muscovy in 1462

Territory Acquired

1462-1533	
1534-1598	
1599-1689	
1690-1725	
1726-1800	
1801-1855	
1856-1876	Territory
1877-1900	Subsequently
1901-1945	Lost

INSET (B) WESTERN FRONTIER

Territory lost after 1913 and never regained

Territory lost after 1913 and regained 1940-45

Territory held throughout the twentieth century

Territory gained in 1945

INSET (C) RUSSIAN TERRITORY IN NORTH AMERICA

Key as Map (A)

C

0   Km   1000

13 and 14 POPULATION DISTRIBUTION AND DENSITY

Population distribution and density are closely interlinked, not least because both are derived from the same data set, namely the number of people recorded within each territorial unit. A population distribution map shows the absolute numbers present in each division, whereas the density map relates those numbers to the land area of each unit. Thus it seems more sensible to treat these two elements together rather than in isolation from each other.

In both cases, the amounts of information which can be given in cartographic form are conditioned by the number, size, shape and arrangement of the administrative units for which data are made available. This situation, which applies to all types of cartography involving the use of statistical data for a set of areal units, presents serious problems in population mapping where the administrative units themselves are to some extent related to the general distribution of population and show a pronounced tendency towards large units in areas of low population density and small ones in zones of population concentration. Nowhere is this more clearly seen than in the USSR. Of the 173 territorial units forming the base for Maps 13 and 14, a mere 20 account for the whole of the East Siberian and Far Eastern regions (46 per cent of the national territory) while the European section of the country (24 per cent of the total) is divided into no fewer than 101 administrative areas. Any density map inevitably shows an average value for each area at the expense of local variation and this averaging effect is particularly marked in Siberia, where the territorial divisions are extremely large.

With an area of 22.4 million sq. km and a (1979) population of 262.4 million, the USSR has an overall population density of 11.7 per sq. km) barely one third of the world average, this low average figure reflecting the presence within the Soviet Union of large areas of territory where the physical environment (Maps 3–9) is basically hostile to human settlement.

A corollary of this situation is the extreme unevenness of the distribution of the Soviet population within the national territory, a feature visible at all levels of spatial resolution. As the Lorenz curve of population concentration attached to Map 14 indicates, three quarters of the total population of the USSR is found within the most densely populated 17 per cent of the land area, while the more densely settled half of the country contains about 96 per cent of the population. This extremely uneven distribution is visible at all levels of spatial aggregation from the four major geographical divisions to the individual administrative units (Table 1).

TABLE 1 AREA and POPULATION

	Area 000 km² (%)	Population 000 (%)	Density per km²
(a) Four geographical divisions			
European USSR	5,455.8 (24.4)	180,266 (68.7)	33.0
Siberia & Far East	12,765.9 (57.0)	27,936 (10.6)	2.2
Kazakhstan & C. Asia	3,994.4 (17.8)	40,165 (15.3)	10.1
Transcaucasia	186.1 (0.8)	14,075 (5.4)	75.6
USSR	22,402.2 (100.0)	262,442 (100.0)	11.7
(b) Sample administrative areas			
Andizhan oblast	4.2 (0.019)	1,349 (0.514)	321.2
Moscow oblast	47.0 (0.210)	14,371 (5.476)	305.8
Taymyr A. Ok	862.1 (3.848)	44 (0.017)	0.05
Evenki A. Ok	767.6 (3.426)	16 (0.006)	0.02

The European USSR, which covers barely a quarter of the national territory, contains two-thirds of the population, while Siberia and the Far East, with 57 per cent of the area, are the homes of only one-tenth of all Soviet citizens. In Kazakhstan and Central Asia, area and population are more closely matched – hence their near-average overall density – though at the level of administrative divisions there are major contrasts. The small but distinctive Transcaucasian region is clearly a zone of population concentration.

Such generalised data, while they form a useful starting point for discussion, are in a sense misleading, since they average out distributions and densities over very large areas and mask strong variations within each region, particularly in the Asiatic section of the country. Maps 13 and 14 give a more detailed picture, showing distribution and density at the level of oblasts or equivalent administrative areas, the smallest units for which the necessary data are available.

The density map (Map 14) shows, by a thickened line, those units with population densities which are above the national average. A main settled zone is clearly visible. This is widest in the west, where it extends from the latitude of Leningrad southwards to the Black Sea and into the North Caucasus and Transcaucasia, and narrows eastwards across the Urals to the Kemerovo oblast (which contains the Kuzbass industrial district). In environmental terms (Maps 5–9) this main settled zone covers the southern fringes of the European tayga, together with the mixed and deciduous forest, wooded steppe and steppe zones and the subtropical lowlands of Transcaucasia. Covering about 20 per cent of the national territory, these areas contain some 72 per cent of the Soviet population – about 189 million people. In general terms, population distribution within this zone is strongly influenced by the agricultural productivity of the land; among heavily rural areas, the highest densities are recorded in climatically favoured districts such as Moldavia which, although only 44 per cent urban, has a density in excess of 100 per sq. km. The highest densities of all, however, are associated with the major urban/industrial concentrations. The Moscow oblast, for example (the largest symbol on the distribution map) has a total population of 14.4 million, 89 per cent of whom are classed as urban, and a density of 306 per sq. km; the three most densely populated oblasts of the Ukraine (Voroshilovgrad, Donetsk and Dnepropetrovsk) have a combined population of 11.6 million (85 per cent urban) and a density of 136 per sq. km. Elsewhere, several oblasts containing major cities form high-density patches on the map – the Kiev and Lvov oblasts for example – though in some cases – Leningrad, Gorkiy, the Urals – the presence of major urban concentrations is masked by the large size of their oblasts.

POPULATION DISTRIBUTION

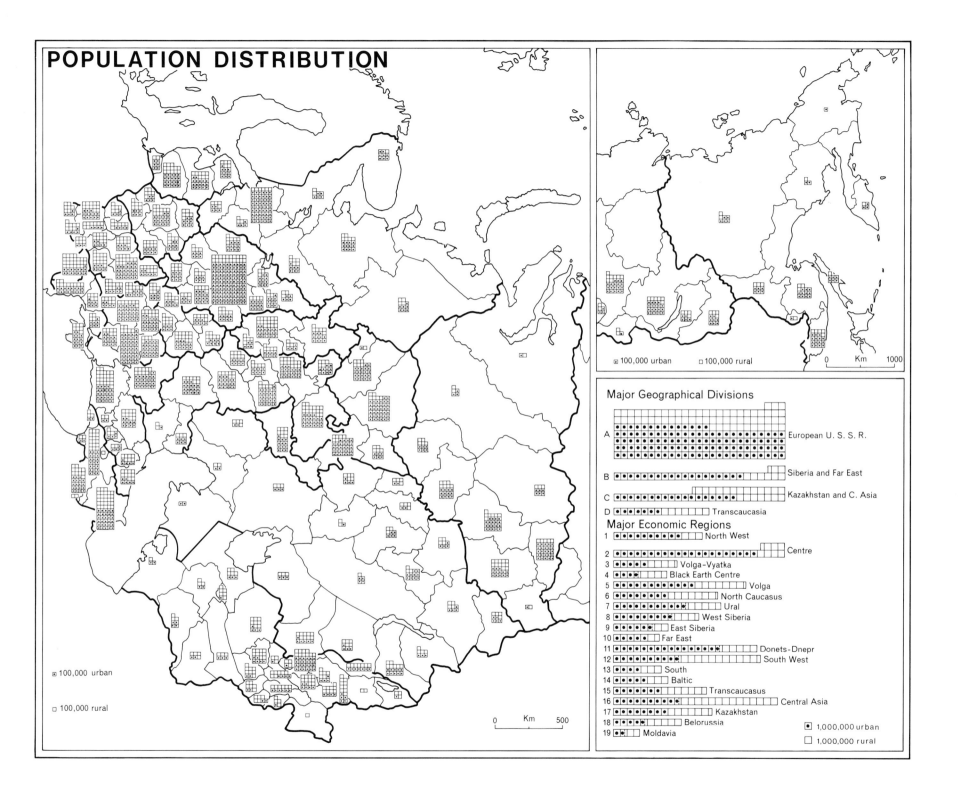

100,000 urban   100,000 rural   0 Km 1000

100,000 urban

100,000 rural

0 Km 500

Major Geographical Divisions

A European U. S. S. R.
B Siberia and Far East
C Kazakhstan and C. Asia
D Transcaucasia

Major Economic Regions

1 North West
2 Centre
3 Volga-Vyatka
4 Black Earth Centre
5 Volga
6 North Caucasus
7 Ural
8 West Siberia
9 East Siberia
10 Far East
11 Donets-Dnepr
12 South West
13 South
14 Baltic
15 Transcaucasus
16 Central Asia
17 Kazakhstan
18 Belorussia
19 Moldavia

1,000,000 urban
1,000,000 rural

The maps show a second, much smaller zone of population concentration in Soviet Central Asia, which includes those areas where irrigated agriculture is especially well developed. The above-average density units in this part of the country have a combined population of nearly 25 million, nearly one tenth of the Soviet total and not far short of the entire population of Siberia and the Far East, within little more than 2 per cent of the land area. Only in the case of the Tashkent oblast is the high density due primarily to the presence of a major city and the situation is essentially one of very high rural densities on very intensively utilised farmland. Particularly striking in this respect are the three oblasts of the Fergana basin – Andizhan, Namangan and Fergana – which have a combined population of 4.1 million and a density of 215 per sq. km, despite being only 32 per cent urban; the rural density alone averages 145 per sq. km, the highest oblast figures in the whole country.

In complete contrast are the vast zones of below-average density which, as a whole, cover about three-quarters of the land area of the USSR but contain only about 15 per cent of the population. Such areas fall clearly into two main categories. By far the largest is the northerly zone, approximately coinciding with the tundra and tayga natural vegetation types, which covers north European USSR, much of Western Siberia and the whole of the East Siberian and Far Eastern regions except for the Maritime Kray. Accounting for about 60 per cent of the land area, this zone contains some 22 million people, little more than 8 per cent of the total. To some extent, the maps are misleading with respect to these areas, which are characterised by particularly large administrative divisions which mask a good deal of internal variation and hide the relatively small areas of population concentration which occur in lowland pockets along the southern edges of the Siberian regions. The general situation in the more northerly parts is typified by the large Yakut ASSR, the biggest administrative unit in the country. With an area of 3.1 million sq. km. – about one-seventh of all Soviet territory – this unit has a population of only 839,000 (0.37 per cent of the total) and an average density of only 0.27 per sq. km).

A second low density zone covers much of the desert and semi-desert sections of Kazakhstan and Central Asia, together with the fringing high mountain districts of the extreme south, where 6 per cent of the population live in 15 per cent of the land area.

Finally, the distribution map (Map 13) distinguishes between the rural and urban populations of each administrative division. Of a total (1979) population of 262.4 million, 163.6 million or 62.3 per cent lived in settlements classed as urban. As already suggested, and as the diagram attached to Map 13 indicate, there are pronounced regional variations in this proportion. Even at the level of the four main geographical divisions, there is a clear distinction between the European USSR and Siberia and the Far East on the one hand, whose populations are 65.5 and 69.6 per cent urban respectively, and Transcaucasia and Kazakhstan and Central Asia on the other, where the urban proportions are well below the national average at 55.4 and 45.5 per cent respectively. Contrasts become more striking at the level of the nineteen Major Economic Regions. Those with urban proportions well above the average include the North-west (79.5 per cent), Centre (78.4 per cent), Donets-Dnepr (75.4 per cent) and Ural (74.4 per cent), while levels of urbanisation well below average are recorded in Moldavia (39.3 per cent), Central Asia (40.7 per cent), the South-west (47.1 per cent), the Black Earth Centre (52.1 per cent), Kazakhstan (53.9 per cent) and the North Caucasus (54.9 per cent). The widest range is, of course, that between individual administrative divisions – from over 90 per cent in the Leningrad oblast to less than 20 per cent in several parts of Central Asia (Naryn oblast, Surkhandarya oblast, Gorno-Badakhshan Autonomous Oblast).

As might be expected, the highest levels of urbanisation are associated with the major industrial zones, with values well above 80 per cent recorded in the Moscow oblast, the Voroshilovgrad, Donetsk and Dnepropetrovsk oblasts of the eastern Ukraine, Sverdlovsk oblast (Urals), Kemerovo oblast (Kuzbass) and the Karaganda oblast of Kazakhstan. Over large areas of the main settled zone, however, particularly in the European sector, the urban proportion is well below the national average in districts where rural populations, though declining, are still large and urban growth has been rather slow. In Moldavia and many of the oblasts of the western Ukraine and over much of Central Asia, more than 60 per cent of the population still live in rural settlements.

A final noteworthy feature is the high level of urbanisation recorded in many of the remote, inhospitable and thinly settled districts of the north, where the possibilities for agriculture are extremely limited and the great bulk of the population lives in ports, mining towns and lumbering centres. Murmansk oblast, for example, is 89 per cent, the Khanty-Mansiy area 78 per cent and the Kamchatka oblast 87 per cent urban.

POPULATION DENSITY

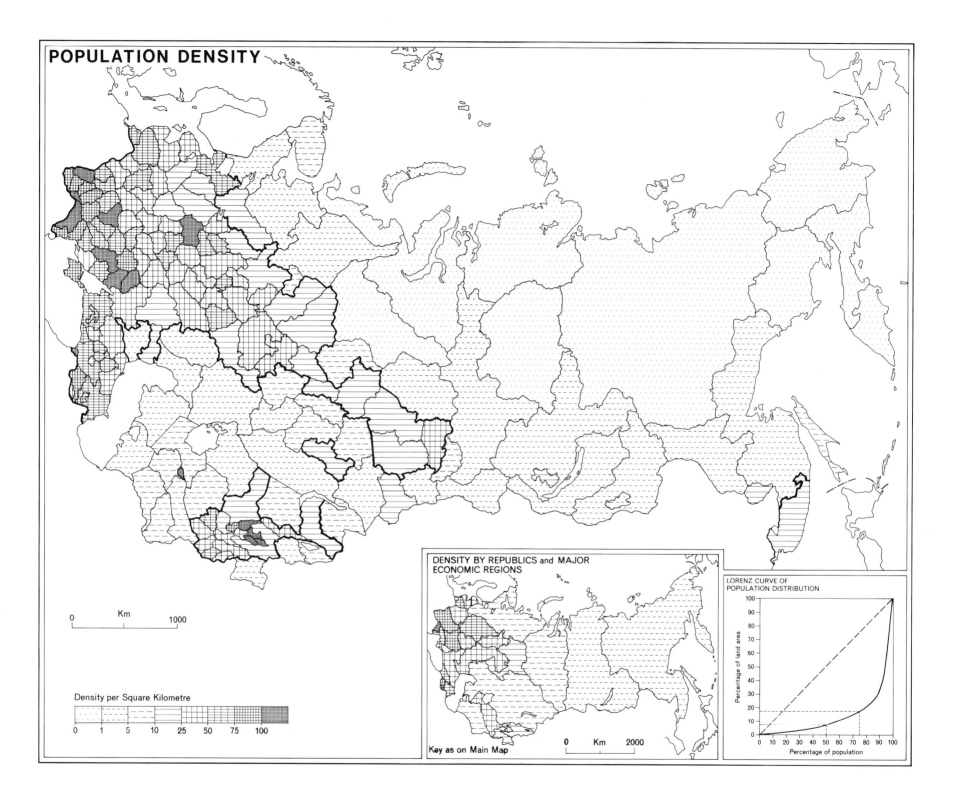

Km
0   1000

Density per Square Kilometre

| | | | | | | | |
0   1   5   10   25   50   75   100

DENSITY BY REPUBLICS and MAJOR ECONOMIC REGIONS

Key as on Main Map

Km
0   2000

LORENZ CURVE OF POPULATION DISTRIBUTION

Percentage of land area

Percentage of population

15 POPULATION CHANGE, 1970–9

In the period since the Revolution, the population of the Soviet Union has risen by more than 60 per cent from 163 million in 1917 to 262 million at the census held in January 1979. Growth would have been a good deal greater had it not been for the effects of two World Wars, the Revolution and Civil War and the forced collectivisation of agriculture, all of which resulted in exceptionally high mortality and a significant reduction in fertility. The growth which has occurred has depended almost entirely on natural increase.

Available information on vital rates is displayed in the graphs which accompany Map 15, where queries indicate a lack of data from the periods of conflict already mentioned. Prior to World War II there was a steady decline in the birth rate, which fell from 45.5 per 1000 in 1913 to 31.2 in 1940. Over the same period the death rate also declined (from 29.1 to 18.0 per 1000), the net result being a fairly steady rate of natural increase around 1.3/1.6 per cent per annum. During the late 1940s and the 1950s, the birth rate fluctuated between 25 and 27 per 1000 while the death rate fell well below the pre-war level to only 7 or 8 per 1000. Thus a natural increase of 1.5 to 1.7 per cent was maintained. During the 1960s, however, there was a pronounced fall in the birth rate to a minimum of 17 per 1000 in 1969 and the death rate edged upwards owing to the ageing of the population. Natural increase fell to 0.9 per cent and has remained about that level during the 1970s.

Map 15 thus refers to a period when population growth in the USSR, though still quite vigorous when compared with most European countries, was slower than in any previous period of peace. Between the censuses of 1970 and 1979, the Soviet population rose from 241.7 to 262.4 million, an increase of 20.7 million or 8.6 per cent. The urban population grew by 27.6 million (20.3 per cent) from 136.0 to 163.6 million, while rural numbers declined by 6.9 million (6.5 per cent) from 105.7 to 98.8 million.

Pronounced regional contrasts are visible at all levels. Among the four major geographical divisions (Table 2) there is a clear distinction between Europe, Siberia and the Far East on one hand and Kazakhstan, Central Asia and Transcaucasia on the other.

TABLE 2 POPULATION CHANGE, 1970–79

	Urban		Rural		Total	
	000	*(%)*	*000*	*(%)*	*000*	*(%)*
European USSR	+18,598	(+18.7)	−9,605	(−13.4)	+ 8,993	(+ 5.3)
Siberia & Far East	+ 3,279	(+20.8)	− 695	(− 7.6)	+ 2,584	(+10.2)
Kazakhstan & C. Asia	+ 4,214	(+30.0)	+3,150	(+16.8)	+ 7,364	(+22.5)
Transcaucasia	+ 1,508	(+23.4)	+ 271	(+ 4.5)	+ 1,779	(+14.5)
Total	+27,599	(+20.3)	−6,879	(− 6.5)	+20,720	(+ 8.6)

In the European sector, total growth significantly below the natural average involved relatively slow urban increase and pronounced rural decline; Siberia and the Far East showed total growth, urban increase and rural decline quite close to the average for the country as a whole. Kazakhstan and Central Asia and, to a lesser degree, Transcaucasia, experienced not only well above average urban growth but also a continuing increase in rural numbers.

The map is designed to identify three major situations. Cross-hatching indicates areas where growth was above the national average and these are enclosed by a thickened boundary; dotted areas are those where the total population actually declined; the intermediate horizontal shadings indicate growth below the national average of 8.6 per cent.

A large area of slow growth or decline covers much of the main settled zone. Throughout practically the whole of this area, rural populations declined – often by more than 20 per cent – and urban growth was relatively slow. Natural increase among the Slav majority has fallen to a very low level and migration movements have occurred not only from rural to urban areas within the zone but also from this zone to other parts of the country. The few areas of above-average increase include Moldavia, with its

relatively high birth rate, Lithuania and a few southerly oblasts (e.g. Crimea) with some net migration gain, and individual oblasts containing major cities – Moscow, Leningrad, Kiev, Minsk, Kuybyshev. The situation in the Centre is characteristic: between 1970 and 1979, the population of this region increased by only 4.8 per cent, involving an urban growth of only 15 per cent and a 21 per cent rural decline.

In complete contrast are the two southerly areas of Transcaucasia and Kazakhstan/Central Asia, where growth was three or four times as rapid as in the European zone. In Transcaucasia, the Georgian Republic joined the slow growth/rural decline category, while in Azerbaydzhan and Armenia high birth rates supported rapid growth and rural numbers continued to increase. In Kazakhstan, northerly steppeland areas showed rather slow growth, including rural decline, but more southerly oblasts were similar to the four Central Asian republics. These experienced rapid growth, both urban and rural, involving a high natural increase rate among the indigenous Moslem groups and a net migration gain from the European regions. Uzbekistan, the most populous of the Central Asian republics, increased its population by 30.4 per cent – a growth rate nearly six times that of the European USSR – from 11.8 to 15.4 million; the urban population grew by 2 million (46.9 per cent) and that of rural areas by 1.6 million (20.6 per cent).

Finally, much of East Siberia and the Far East, together with the northern parts of European USSR and West Siberia, experienced very rapid growth, though the numbers involved were generally small. The highest growth rates in the country were associated with specific economic developments. Thus the exploitation of oil and gas involved an influx of people into the Tyumen oblast (34 per cent increase) and especially into the Khanty-Mansiy and Yamalo-Nenets Auton- omous Okrugs: the population of these two areas more than doubled — but only from 351,000 to 727,000.

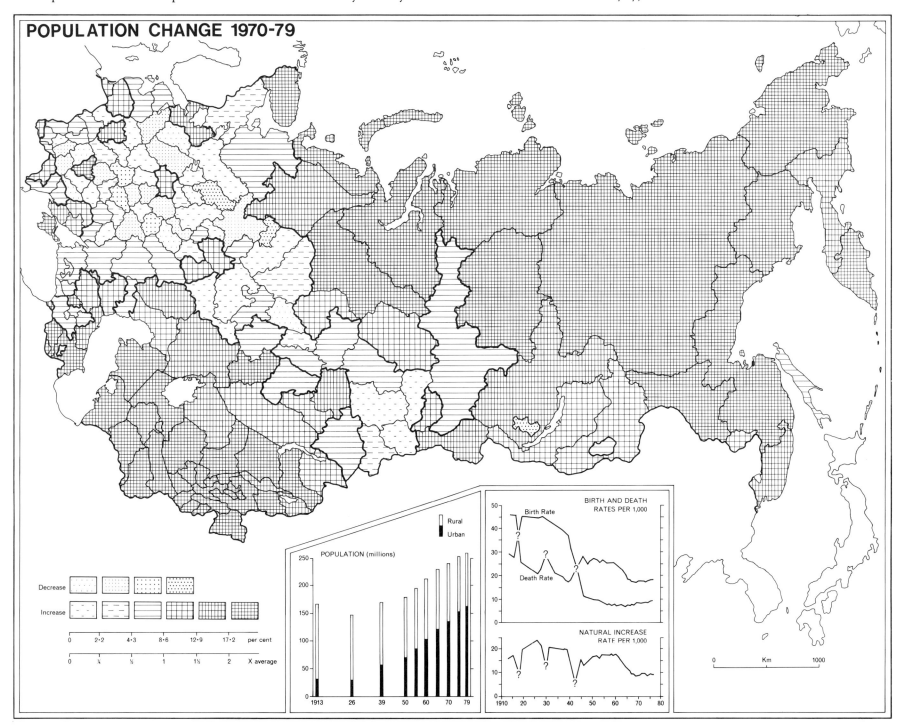

POPULATION CHANGE 1970-79

Decrease

Increase

| 0 | 2·2 | 4·3 | 8·6 | 12·9 | 17·2 | per cent |

| 0 | ¼ | ½ | 1 | 1½ | 2 | X average |

POPULATION (millions)

☐ Rural
■ Urban

250

200

150

100

50

0

1913 26 39 50 60 70 79

BIRTH AND DEATH RATES PER 1,000

Birth Rate

Death Rate

NATURAL INCREASE RATE PER 1,000

0 Km 1000

33

16 NATIONALITIES

The expansion of the Russian Empire (Map 12) brought within its boundaries a great variety of peoples differing widely in race, language, religion, culture, historical background and mode of life, and the multi-national character of the modern USSR is recognised in its internal political geography (Map 10). The Soviet census of 1979 recognised no fewer than 92 separate national groups within the Soviet population.

Although, in a minority of cases, the individual's national and linguistic affiliations may be different, the defining attribute of a nationality is its possession of a distinctive language: thus the classification of nationalities is best carried out on a linguistic basis. Full details of all 92 national groups cannot be provided in the space available and attention must be focussed on the larger groups. Table 3 shows the allocation of nationalities to size groups, while Table 4 lists the 37 groups with populations in excess of 250,000. These 37 groups constitute 98.4 per cent of the Soviet population. Map 16 shows the areas where these groups (along with an additional 20 smaller groups) constitute a major element in the local population. Solid shadings show the main settlement areas of each group; broken shadings are thinly settled areas.

The Soviet Union contains representatives of two of the world's major language families, the Indo-European (Maps 16A, B), spoken by nearly 200 million people, and the Ural-Altaic (Maps 16D, E, F) spoken by about 35 million; about 5 million people living in the Caucasus speak languages which are not allocated to either of these major categories.

Indo-European languages are spoken over a wide area of Eurasia from the Bay of Bengal to the Atlantic. In Europe, the major sub-divisions are the Celtic, Teutonic, Romance and Slav and all but the first are represented in the USSR Overwhelmingly predominant are the Slavs (Map 16A) – Russians (1), Ukrainians (2) and Belorussians (3) – who, together with the small numbers of Poles (4) and Bulgars (5), constitute three-quarters of the Soviet population. Slavs are dominant over most of the European plain and, having migrated eastwards in large numbers, are a majority in most of the settled districts of Siberia.

Non-Slav Indo-European groups (Map 16B) are much smaller. A Baltic sub-group comprises the Lithuanians (6) and Latvians (7), while the Moldavians (8) speak a Romance language akin to Romanian. The Indo-Iranian sub-group includes the Tadzhiks (11), a Moslem people in Central Asia, and several small nationalities in the Caucasus. The Armenians (15) are highly distinctive, with their own unique script and form of Christianity, but are generally classed as Indo-European. The Germans (9) are mainly descendants of eighteenth and nineteenth-century immigrants, who lived in the European USSR prior to World War II but were dispersed to western Siberia and Kazakhstan during the war. Finally the Jews (16), whose numbers have recently declined owing to emigration, are found mainly in the European cities; only very small numbers have settled in the Far Eastern 'homeland' established for them.

Ural-Altaic languages are spoken by only about 15 per cent of the Soviet population but these include a very large number of national groups; at least five major sub-divisions have been identified.

Finno-Ugrian groups (Map 16D) are found mainly in the northern half of the European USSR and western Siberia, where they were established before the arrival of Russian settlers. The most advanced of these are the Estonians (25) of the Baltic and a number of groups between the Volga and the Urals – the Udmurts (30), Mariy (31) and Mordov (32). The Komi (28), Komi-Permyak (29), Mansiy (33), Khanty (34) and Nentsy (35) are more primitive.

Turkic speakers (Map 16E) number nearly 30 million. They are dominant throughout Soviet Central Asia, where the major groups are the Uzbeks (41), Kazakhs (36), Turkmen (43) and Kirgiz (40). There are also important Turkic-speaking groups – the Tatars (38) and Bashkirs (39) – in the Volga-Ural zone and western Siberia. The Azerbaydzhanis (44) and numerous smaller groups in the Caucasus are also Turkic. Smaller Turkic groups – the Khakass (48), Altays (47) and Tuvinians (49) – occur in southern Siberia and a large area of north-east Siberia is occupied mainly by the Yakuts (50).

Map 16F shows the distribution of the main groups in the Mongolian, Tungus-Manchurian and Paleo-Asiatic subdivisions of the Ural-Altaic and indicates the considerable ethnic diversity of eastern Siberia and the Far East. The Kalmyks (53) are a special case – a Mongolian group which migrated westwards in the seventeenth century.

TABLE 3 NATIONALITIES in 1979 by SIZE GROUPS

Size group (millions)	No.	Population	
		000	%
over 10.0	3	192,200	73.23
5.0 – 10.0	4	27,813	10.60
2.5 – 5.0	5	16,439	6.26
1.0 – 2.5	10	15,605	5.95
0.5 – 1.0	4	2,634	1.00
0.25 – 0.5	11	3,590	1.37
0.1 – 0.25	15	2,234	0.85
below 0.1	40	750	0.29
unclassifed	–	1,175	0.45
Total	92	262,440	100.00

TABLE 4 THE 37 LARGER NATIONALITIES (Those with more than 0.1 per cent of the total population, i.e. 262,440) Populations in thousands; percentage of Soviet total in brackets.

Russians 137,397 (52.35); Ukrainians 42,347 (16.14); Uzbeks 12,456 (4.75); Belorussians 9,463 (3.61); Kazakhs 6,556 (2.50); Tatars 6,137 (2.41); Azerbaydzhanis 5,477 (2.09); Armenians 4,151 (1.58); Georgians 3,571 (1.36); Moldavians 2,968 (1.13); Tadzhiks 2,898 (1.10); Lithuanians 2,851 (1.09); Turkmen 2,028 (0.77); Germans 1,936 (0.74); Kirgiz 1,906 (0.73); Jews 1,811 (0.69); Chuvash 1,751 (0.67); Latvians 1,439 (0.55); Bashkirs 1,371 (0.52); Mordovs 1,192 (0.45); Poles 1,151 (0.44); Estonians 1,020 (0.39); Chechens 756 (0.29); Udmurts 714 (0.27); Mariy 622 (0.24); Osetins 542 (0.21); Avars 483 (0.18); Koreans 389 (0.15); Lezhgins 383 (0.15); Bulgars 361 (0.14); Greeks 344 (0.13); Yakuts 328 (0.12); Komi 327 (0.12); Kabardins 322 (0.12); Karakalpaks 303 (0.12); Dargins 287 (0.11)

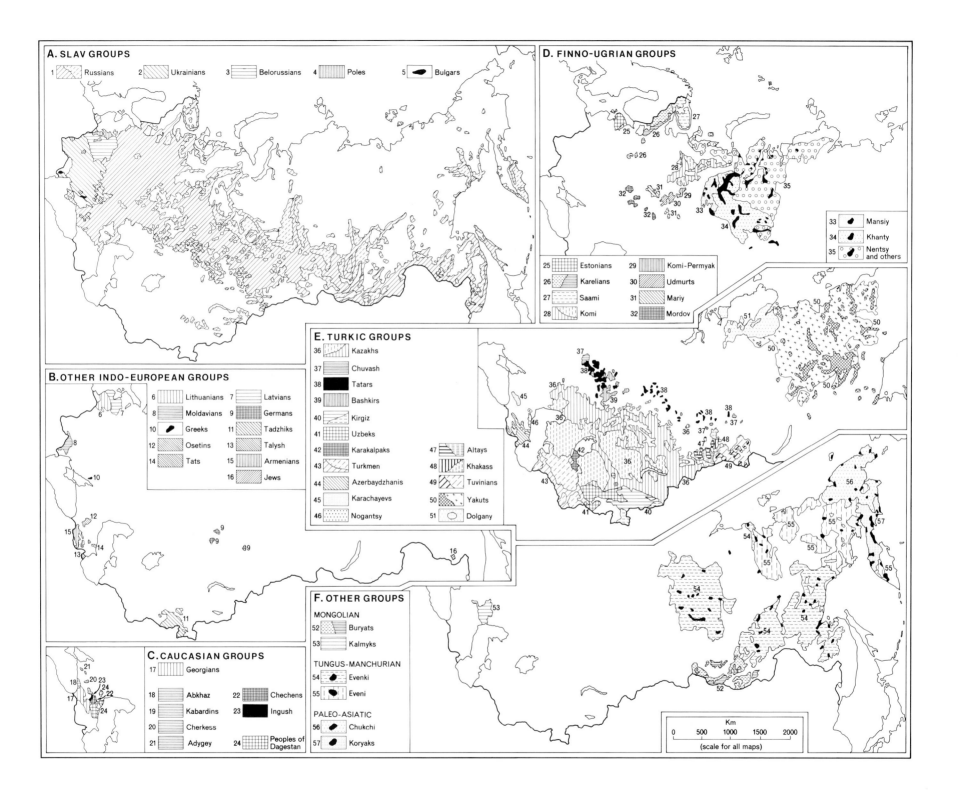

A. SLAV GROUPS
1 Russians 2 Ukrainians 3 Belorussians 4 Poles 5 Bulgars

B. OTHER INDO-EUROPEAN GROUPS
6 Lithuanians 7 Latvians
8 Moldavians 9 Germans
10 Greeks 11 Tadzhiks
12 Osetins 13 Talysh
14 Tats 15 Armenians
16 Jews

C. CAUCASIAN GROUPS
17 Georgians
18 Abkhaz 22 Chechens
19 Kabardins 23 Ingush
20 Cherkess
21 Adygey 24 Peoples of Dagestan

D. FINNO-UGRIAN GROUPS
25 Estonians 29 Komi-Permyak
26 Karelians 30 Udmurts
27 Saami 31 Mariy
28 Komi 32 Mordov

33 Mansiy
34 Khanty
35 Nentsy and others

E. TURKIC GROUPS
36 Kazakhs
37 Chuvash
38 Tatars
39 Bashkirs
40 Kirgiz
41 Uzbeks
42 Karakalpaks 47 Altays
43 Turkmen 48 Khakass
44 Azerbaydzhanis 49 Tuvinians
45 Karachayevs 50 Yakuts
46 Nogantsy 51 Dolgany

F. OTHER GROUPS
MONGOLIAN
52 Buryats
53 Kalmyks

TUNGUS-MANCHURIAN
54 Evenki
55 Eveni

PALEO-ASIATIC
56 Chukchi
57 Koryaks

Km
0 500 1000 1500 2000
(scale for all maps)

35

17 TOWNS

The rapid and widespread industrialisation of the USSR in the period since the Revolution has been accompanied by massive urbanisation and rapid growth of the urban population both in absolute numbers and as a proportion of the total. Whereas in 1926 only 17.9 per cent of the population lived in settlements classified as urban, by 1979 the proportion had risen to 62.3 per cent. Over the same period, the number of town-dwellers multiplied six-fold, rising from 26.3 to 163.6 million. Much of this growth has occurred in the period since World War II. Although in the inter-war years the *rate* of urbanisation was extremely high, by 1939 there were still only 60.4 million town-dwellers, just below one third of the total population, and the landmark of 50 per cent urban was not reached until 1961. Even today, the level of urbanisation is somewhat low when compared with those of most developed industrial countries.

1926 only about one third of all town-dwellers were in cities with over 100,000 inhabitants, this proportion reached 47 per cent in 1939 and 61 per cent in 1979, with 32 per cent living in major centres with more than 500,000 inhabitants. In the past 20 years, the number of people living in small towns with fewer than 50,000 inhabitants has increased by about 20 per cent, but the number living in cities of over 500,000 has doubled.

Map 17 shows the location of approximately 500 urban settlements – those with over 50,000 inhabitants – of which 108 with more than 250,000 are named on the map. The USSR now has eighteen cities with populations in excess of one million. Only two of these – Moscow (8,011)* and Leningrad (4,588), the current and pre-Revolutionary capitals respectively – have been million cities throughout the Soviet period, and in size and importance they stand head and shoulders above all other Soviet cities. They were joined by Kiyev (now 2,144), capital of the Ukraine, in the late 1950s, but the remaining 15 have achieved this level only during the last twenty years. Tashkent (1,779), Baku (1,550), Kharkov (1,444), Gorkiy (1,344) and Minsk (1,276) passed the million mark before the 1970

census; all are republican and/or major regional capitals as well as being important industrial centres. The remaining ten have all emerged during the 1970s; these include leading industrial and/or regional centres – Novosibirsk (1,312), Kuybyshev (1,216), Sverdlovsk (1,211), Dnepropetrovsk (1,066), Odessa (1,046), Chelyabinsk (1,031) – some of which are also capitals of smaller republics – Tbilisi (1,066), Yerevan (1,019).

Within the 500,000–1 million category, a number of centres with similar functions are fast approaching the million mark, notably Perm (999), Kazan (993), Ufa (969), Rostov (934), Volgograd (929) and Alma-Ata (910), the capital of the Kazakh republic. The remaining 90 cities with more than 250,000 inhabitants are for the most part oblast or kray capitals, with important administrative functions in addition to their industrial activities. The 160-odd towns in the size range 100–250,000 are a good deal more varied in character. This group includes adminstrative and marketing centres in the more rural parts of the country such as the western Ukraine, the North Caucasus region and Central Asia, second-rank industrial towns in highly urbanised areas like the Centre, Donbass, Urals and Kuzbass, and isolated centres of extractive industry such as Norilsk in Central Siberia, Rudnyy in north-western Kazakhstan and Guryev on the north coast of the Caspian. Small towns with fewer than 50,000 inhabitants are commonly satellites of their larger neighbours, as around Leningrad, Moscow and Gorkiy and in the Donbass, but are also common in some predominantly rural areas such as Belorussia.

The overall pattern of urban centres depicted on Map 17 bears an obvious relationship to the general distribution of population, with close spacing of towns in the more densely populated areas and large distances between centres in the thinly settled parts of the country. Particularly striking are the clusters of towns in the Moscow basin, eastern Ukraine, central Urals and Kuzbass, the evenly-spaced line of major cities along the Volga and the string of towns along the Trans-Siberian railway.

TABLE 5 URBAN POPULATION BY TOWN SIZE 1926–79

	1926		1939		1959		1970		1979	
	mill	(%)	mill	(%)	mill	(%)	mill	(%)	mill	(%)
Over 1 million	3.6	(13.7)	7.6	(12.6)	10.5	(10.5)	20.9	(15.4)	33.1	(20.2)
500,001 – 1 million	0.5	(1.9)	5.2	(8.6)	13.7	(13.7)	16.3	(12.0)	18.6	(11.4)
250,001 – 500,000	2.2	(8.4)	6.7	(11.1)	11.4	(11.4)	16.9	(12.4)	22.0	(13.4)
100,001 – 250,000	3.2	(12.2)	8.9	(14.7)	13.0	(13.0)	21.4	(15.7)	26.2	(16.0)
50,000 – 100,000	4.1	(15.6)	7.0	(11.6)	10.9	(10.9)	13.1	(9.6)	63.7	(38.9)
below 50,000	12.7	(48.3)	25.0	(41.4)	40.5	(40.5)	47.4	(34.9)		
Total urban	26.3	(100.0)	60.4	(100.0)	100.0	(100.0)	136.0	(100.0)	163.6	(100.0)
Urban	26.3	(17.9)	60.4	(31.7)	100.0	(47.9)	136.0	(56.3)	163.6	(62.3)
Rural	120.7	(82.1)	130.3	(68.3)	108.8	(52.1)	105.7	(43.7)	98.8	(37.7)
Total	147.0	(100.0)	190.7	(100.0)	208.8	(100.0)	241.7	(100.0)	262.4	(100.0)

As the table shows, the increase in total urban population has been accompanied by a rise in the proportion living in the larger cities. Whereas in

* A figure in brackets indicates the city population in thousands at the 1979 census

Inevitably, the scale of this map has involved considerable compression and it may not always be possible to locate or identify specific towns. The same set of towns is, however, mapped in more detail and in conjunction with other features on the regional maps (41–49) on pages 88–105.

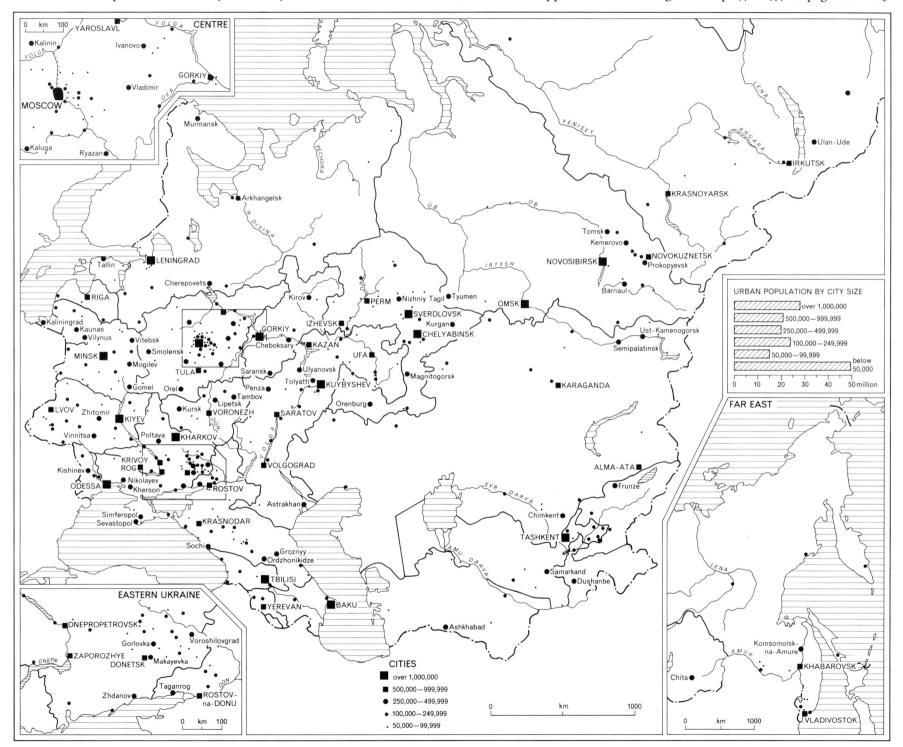

CENTRE

URBAN POPULATION BY CITY SIZE
over 1,000,000
500,000–999,999
250,000–499,999
100,000–249,999
50,000–99,999
below 50,000
0   10   20   30   40   50 million

FAR EAST

EASTERN UKRAINE

CITIES
■ over 1,000,000
■ 500,000–999,999
● 250,000–499,999
• 100,000–249,999
· 50,000–99,999

0   km   1000

PART 3 ECONOMIC GEOGRAPHY

18 and 19 CROP DISTRIBUTIONS

These twelve small maps show, in a highly generalised manner, the distribution of areas engaged in the production of seventeen different crops. In the majority of cases, a denser shading is used to indicate the areas of most intensive production of the named crop. The distributions are of two kinds. In several instances (e.g. rye, oats, barley, potatoes) the crop is grown to some extent in almost every area; in others (e.g. cotton, hemp, maize, tobacco, tea) the narrower climatic requirement of the crop confine its production to restricted sections of Soviet territory.

Map 18A Agricultural Areas

This shows the general distribution of agricultural land of all types. Such land covers about a quarter of the total land area of the USSR (for details see p. 46) and about 40 per cent of the agricultural land is used for the production of crops. The allocation of the sown area among individual crops is given in the accompanying table. The distribution of agricultural land is related primarily to the environmental factors discussed on

TABLE 6 MAJOR CROPS: AREA AND PERCENTAGE OF TOTAL SOWN (1979)

	000 ha	(%)		000 ha	(%)
Cereals	121,400	(55.9)	Industrial		
Wheat	57,700	(26.6)	crops	14,400	(6.6)
Barley	37,000	(17.0)	Sugar beet	3,740	(1.7)
Oats	12,200	(5.6)	Cotton	3,090	(1.4)
Rye	6,500	(3.0)	Flax	1,050	(0.5)
Millet	2,800	(1.3)	Hemp	120	(n.)
Maize	2,700	(1.2)	Oilseeds	5,810	(2.7)
Buckwheat	1,700	(0.8)	others	590	(0.3)
Rice	600	(0.3)	Fodder crops	67,300	(31.0)
Others	200	(0.1)	Sown grasses	4,200	(20.3)
Legumes	5,000	(2.3)	Maize	6,800	(7.7)
Potatoes	7,000	(3.2)	Root crops	3,800	(0.8)
Vegetables	1,700	(0.8)	others	4,500	(2.2)
Fruit	500	(0.2)	Total Sown	217,300	(100.0)

pages 4–19. It is virtually continuous throughout the main settled zone of the European USSR, embracing the Centre, Black Earth Centre, Volga and North Caucasus regions and the Baltic, Belorussian, Moldavian and Ukrainian republics, though even here there are numerous patches of swamp, forest and other non-agricultural land too small to be shown at this scale. North of latitude 56°N, however, agricultural land is confined to major valleys and pockets around the larger settlements. A broad belt of agricultural land, roughly coinciding with the steppe zone, extends across northern Kazakhstan and the southern part of West Siberia and there are important pockets in the southerly lowland basins of East Siberia and the Far East, but over most of these eastern regions it occurs only along the main valleys, where intensity of land use is in any case low. The desert areas of Kazakhstan and Central Asia, together with the southern mountain districts, are blank on the map; between these zones are pockets of intensive agricultural land use based mainly on irrigation.

Map 18B Wheat

Cereals occupy well over half the sown area of the USSR and, of these, wheat is by far the most important, accounting for nearly half the cereal area. 70 per cent of the wheat is spring sown; winter varieties are produced on a significant scale only in the western Ukraine and Moldavia. As the map indicates, wheat is grown in nearly all agricultural areas, the main exceptions being northerly districts, where the growing season is too short, certain ill-drained sections of the forest zone such as the Polesye on the Belorussia/Ukraine border, and some of the southern districts of intensive irrigated farming. The highest proportions of wheatland, however, are to be found in the steppe belt. Wheat occupies 50 per cent or more of the sown area in Moldavia, the Ukraine, North Caucasus and Kazakhstan and 35–40 per cent in the three Siberian regions, but only about 3 per cent in the North-west.

Map 18C Barley

The general distribution of barley production,

which occupies about 30 per cent of the cereal area, is similar to that of wheat, in that this crop is to be found in nearly all agricultural areas. Indeed, the range of conditions under which barley is grown is appreciably wider than that for wheat; barley can withstand both the shorter growing season of northern agricultural areas and the drier conditions of the desert margin.

Map 18D Rye

Like the other cereals, rye is also grown in most agricultural areas, though it occupies only 5 per cent of the cereal land and is virtually absent from Central Asia and Transcaucasia. Rye is the traditional cereal crop of the European forest zone and is grown mainly in the northern half of the main agricultural belt, occupying 10 per cent or more of the total sown area in the Centre, Volga-Vyatka and North-west regions and in the Baltic and Belorussian republics. The area devoted to this crop has steadily diminished and is now less than half that recorded in the 1950s; much of the land formerly under rye is now used for sown grasses and other fodder crops.

Map 18E Oats

This is another ubiquitous crop grown in all agricultural areas. It is most important in the northern part of the European forest zone. The area of land under oats has increased rapidly in recent years, reflecting the much greater attention now being paid to the livestock side of agriculture.

Map 18F Maize and Rice

The production of maize in the USSR has had a chequered career over the past few decades. In the 1950s, the maize area (7 mill ha) was about a third of its present size; in the early 1960s, as a result of Khrushchev's agricultural policies, it shot up to nearly 40 mill ha, but in the late 1960s it declined to its present level of about 19 mill ha. Khrushchev's attempts to achieve large-scale production of maize grain were a failure, since climatic conditions in most areas do not favour

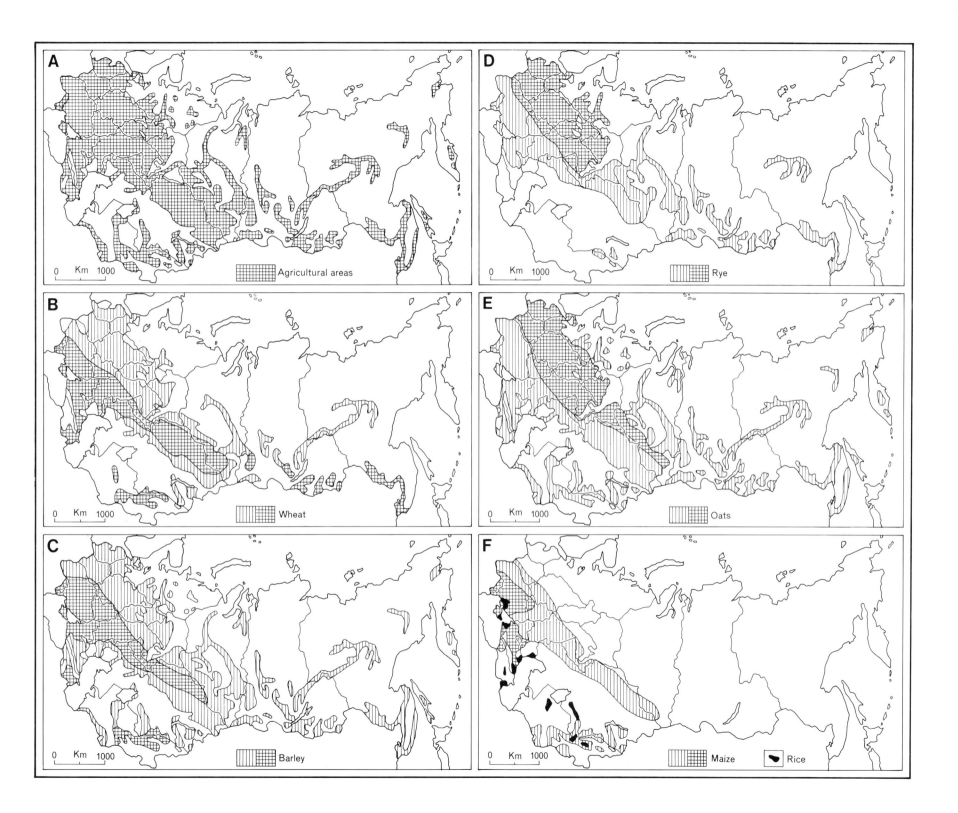

41

the ripening of the crop. The **area** devoted to this purpose (2.7 mill ha) is now much the same as in the 1950s. The use of maize for silage (16.8 mill ha), however, has become a major element in Soviet livestock farming. The crop is produced mainly in areas which, by Soviet standards, are relatively warm and moist, namely Moldavia, the western Ukraine and the North Caucasus.

With only 600,000 hectares of rice land (0.5 per cent of the cereal area), the USSR is a minor producer. The crop is grown under irrigation in Central Asia, in pockets along the western side of the Caspian and in the Kuban (North Caucasus). A more recent development has been the cultivation of rice in the southern Ukraine, where irrigation water is available from barrages along the Dnepr (see p. 92).

Map 19A Potatoes

These occupy about 7 million ha and are another crop grown to some extent in practically all agricultural areas, providing a source of industrial alcohol as well as human and livestock feed. Like rye and oats, potatoes are most important in the northern part of the European forest zone; at least 75 per cent of the area devoted to this crop is in the more densely shaded areas on the map. Potatoes account for 13 per cent of the sown area in Belorussia and 8 per cent in the Centre, which together produce a quarter of the crop.

The production of *industrial crops* has risen steadily since the Revolution and the area devoted to them has increased by some 10 million hectares.

Map 19B Sugar Beet

The area under sugar beet has greatly increased during the Soviet period, rising from less than 700,000 ha at the time of the Revolution to the current 3,740,000 ha. The crop is grown mainly in rotation with cereals, particularly wheat, and is thus found mainly in the moister, north-western part of the steppe and in the wooded steppe, though it is also grown on a small scale in the Baltic republics, Central Asia and the Far East.

50 per cent of the land devoted to sugar beet is in Moldavia and the Ukraine, 20 per cent in the Black Earth Centre – where it occupies more than 5 per cent of the sown area – and a further 17 per cent in the Centre, Volga and North Caucasus regions.

Map 19C Sunflower

This crop, which covers three-quarters of the land devoted to oilseeds, has features similar to those just described for sugar beet and forms another element in the agricultural diversification achieved during the Soviet period. The area under sunflower has multiplied more than fourfold and it is grown mainly in the steppe, generally to the south of the main sugar beet districts. Nearly a quarter of the land under sunflower is found in the North Caucasus region, where it occupies 7 per cent of the sown area, about 40 per cent in the Ukraine and Moldavia and another 30 per cent in the Black Earth Centre and Volga regions. Production elsewhere is insignificant.

Map 19D Flax and Cotton

These two main fibre crops have completely different distribution patterns. Flax, like rye, is a traditional crop of the cool, relatively damp European forest zone and, also like rye, the area used in its production has recently declined, mainly in favour of fodder crops. Flax occupies 6 per cent of the small sown area of the North-western region and about 3 per cent in Belorussia and the Centre. These three regions together have two-thirds of all the flax cultivation in the USSR.

Cotton is a speciality of Central Asia (including southern Kazakhstan), which has 93 per cent of all Soviet cotton land, and the Transcaucasus. In the irrigated districts of these two regions, especially the former, cotton-growing reaches a level of specialisation rarely encountered in Soviet agriculture. 41 per cent of the sown area of the four Central Asian republics is occupied by this one crop, the proportion rising to 47 per cent in Uzbekistan, the chief producer, and 58 per cent in Turkmeniya.

19E Hemp and Tobacco

These two crops are shown on a single map as a matter of convenience: they have different distributions, both within the more southerly parts of the USSR.

Hemp, which occupies only 0.1 per cent of the total sewn area, is grown mainly in the wooded steppe zone, where moisture is quite plentiful and summers are warm; the main concentration is in the Black Earth Centre. The crop is also grown under irrigation in Central Asia.

Tobacco requires high summer temperatures and plentiful moisture: the main tobacco-producing districts are in the western Ukraine and Moldavia, and along the Black Sea coasts of the Crimea, North Caucasus and Georgia. Tobacco is also grown under irrigation in the Transcaucasus and the Central Asian republics.

Map 19F Vines, Tea, Citrus, Soya

Viticulture is a specialism of southerly regions. The USSR has about 1.2 mill ha of vineyards, of which 24 per cent are in the small Moldavian republic alone, 36 per cent in the Ukraine and North Caucasus, 28 per cent in the Transcaucasus and 10 per cent in Central Asia.

The production of *tea* and *citrus fruits* in the Colchis and Talysh lowlands of Transcaucasia reflects the very special climatic conditions of those regions.

Soya bean production is a unique feature of the Amur-Ussuri lowlands in the Far East.

No attempt has been made to map the areas of *fodder crop* production since, at this scale of generalisation, these would include all the agricultural areas shown in Map 18A. Such crops now cover about 31 per cent of the sown area, compared with barely 3 per cent at the time of the Revolution. Sown grasses are dominant (20 per cent of the sown area), followed by maize for silage (8 per cent), leguminous fodders (2 per cent) and roots (less than one per cent). The highest proportion of land under fodders occurs in the cooler, damper regions, such as the North-west (53 per cent), Baltic (45 per cent), Belorussia (36 per cent) and Centre (35 per cent), the lowest in the Central Asian republics (22 per cent).

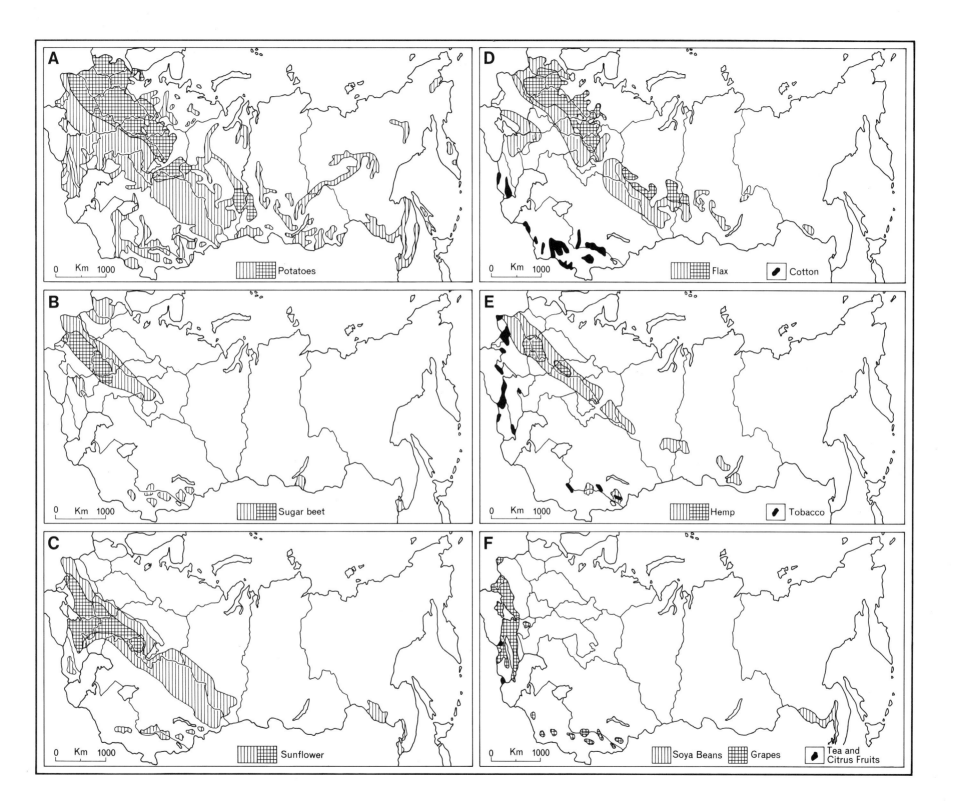

43

20 AGRICULTURAL REGIONS

The varied environmental conditions within the USSR and the processes of agricultural development combine to produce a complex pattern of spatial variations in agricultural practices and types of agricultural production. Four agricultural zones are distinguished (their boundaries indicated by thickened lines on the map); these are divided into a total of ten agricultural regions, some of which are further subdivided to indicate internal variations.

Zone A: Northern Areas of Low Agricultural Value

Throughout this very large zone, which covers the tundra and most of the tayga vegetation belts and occupies at least half the territory of the USSR, agriculture is of minimal significance. The tundra, together with the northern tayga, constitute Region 1 (reindeer herding, hunting and coastal fishing), where agriculture is virtually non-existent. In Region 2 (forest exploitation with small-scale agriculture) small, scattered areas of farmland are devoted to cattle and sheep grazing and the production of limited quantities of rye, oats, barley, potatoes and fodder crops. A rather higher level of agricultural activity occurs in the valleys of the middle Lena and its tributaries (Region 2a); cereals include some spring wheat as well as rye, oats and barley and there is some vegetable and fodder crop production, but the emphasis is on cattle rearing, mainly for beef.

Zone B: The Main Agricultural Belt

This extends from the western frontier to Lake Baykal, coinciding with the mixed and deciduous forest, wooded steppe and steppe environmental zones, and also includes a number of lowland pockets further east. Zone B is the agricultural heartland of the USSR and produces a wide range of crops and livestock under a variety of mixed farming systems. Summer temperatures increase and the growing season lengthens towards the south, while the amount and reliability of precipitation decline from north-west to south-east and, largely for historical reasons, the density of settlement and the intensity of land use fall off east of the Volga and Urals. Three main agricultural regions can be identified.

Region 3 (cereals, flax, dairying) occupies the southern part of the European tayga and the northern section of the mixed and deciduous forest belt. The podsolic soils are rather poor and waterlogging is often a problem but these difficulties can be overcome by the addition of lime and fertiliser and by drainage. The traditional cereal crop, rye, though still important, is now being replaced by other cereals. Oats and sown grasses support large numbers of dairy cattle and there are extensive, rather low grade, pasture lands. Flax is another traditional but contracting crop. Potatoes are a major element, and a striking feature of recent years has been a big increase in fodder crop production.

Region 4 (arable and livestock) covers the western part of the wooded steppe belt and the southern section of the mixed and deciduous forest. It division into western (4a) and eastern (4b) sections denotes less intensive activity in the latter. The crop range is a good deal wider than in region 3. Hemp replaces flax as the main fibre crop, and sugar beet is widespread, especially in the west where it is grown in rotation with wheat. Wheat is the main cereal in the south of the region, where maize is also grown, but in the north it is secondary to barley and oats. Potatoes are widely grown, together with other root crops for stock feeding, particularly of dairy cattle.

Region 5 (cereals and livestock) derives its unity from its coincidence with the steppe vegetation belt and its extreme importance to the Soviet agricultural economy. A major subdivision occurs between the more intensively farmed western section (5a) and the eastern steppelands (5b) beyond the Volga where farming is more extensive. Wheat is the dominant crop throughout the region, supplemented by rye and oats in the north-west, maize in the warm, moist south-west and barley and millet in the dry south-east. During the nineteenth century, much of the European steppe grew little else but wheat; during the twentieth century arable farming has become more diversified with larger areas devoted to sugar-beet, sunflower, root crops and sown grasses, and this has supported a major expansion of the livestock side. East of the Volga there is still a very strong emphasis on wheat production, though this area, too, is becoming more mixed. The eastern steppelands were the scene of a massive expansion of the agricultural area when, during the 1950s, some 40 million hectares were brought into cultivation under the 'Virgin Lands' scheme.

Pockets of wooded steppe and steppe, and thus of cereal and livestock farming, occur further east along the southern edge of Siberia, while in the Far East lowlands (5c) rice and maize are grown as well as wheat and barley, there is much fodder-crop production and soya beans are a major item.

Zone C: Southern Areas of Low Agricultural Value

These are of two types, in both of which livestock are dominant. Region 6 (desert and semi-desert stock-rearing) is traditionally the home of nomadic sheep-herders but, during the Soviet period, true nomadism has virtually disappeared. In the more favoured areas, fodder crops are grown to supplement the scanty natural grazing, and permanent settlements have been established whence herdsmen set out with their flocks on the annual migration circuit. Region 7 (mountain stock-rearing) includes the high Caucasus, the mountains of Central Asia and the Sayan Mountains of southern Siberia. In these districts, the limited arable land is devoted mainly to fodder crops and there is transhumance between low-level winter and high-level summer pastures.

Zone D: Southern Areas of High Agricultural Value

Region 8 (horticulture, viticulture and tobacco production) includes a variety of areas of intensive cultivation benefiting from special local conditions. In Moldavia, vineyards are widespread and fruit, vegetables and tobacco are produced in large quantities. The southern Crimea also is noted for its fruit, tobacco and vines, as are the Caucasian areas in this category. Along the lower Volga valley there is a broad ribbon of irrigated fruit and vegetable production. Region 9 (subtropical crops) embraces the lowlands of humid western Transcaucasia and the small Talysh lowland on the Caspian. Heavy rainfall and high temperatures permit the cultivation of such specialities as citrus fruits, tung nuts (a source of vegetable oil), tobacco, vegetables, maize and tea.

Finally Region 10 (irrigated agriculture) occurs in eastern Transcaucasia and in the Central Asian republics in a hill-foot zone between the mountains and the desert. Cotton is by far the most important crop but these areas also produce large quantities of lucerne, rice, sugar-beet, hemp, tobacco, vines and fruit.

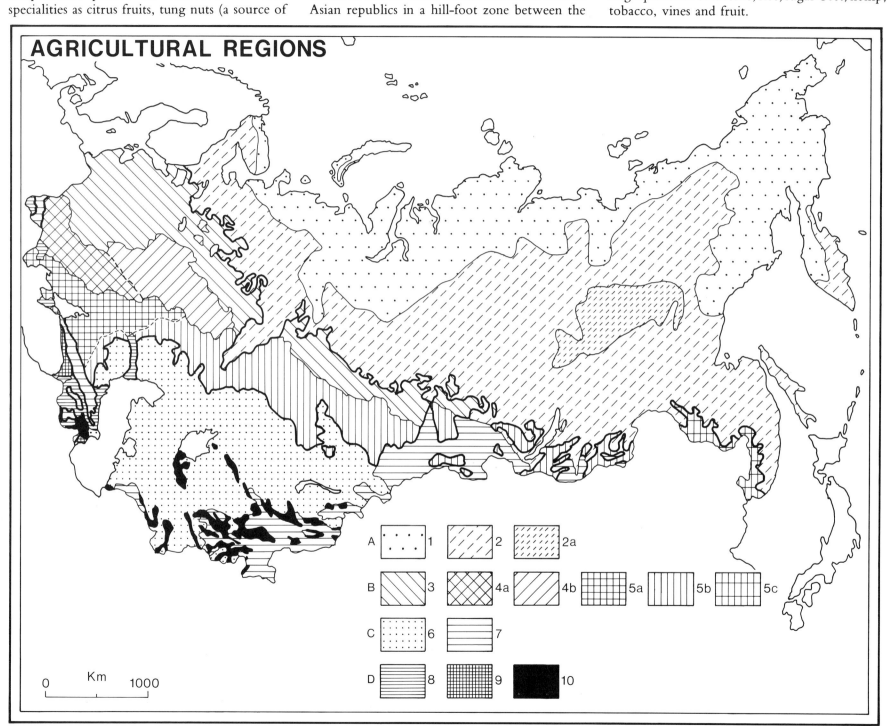

AGRICULTURAL REGIONS

A ⋯ 1   ╱╱ 2   ╱ 2a

B ╱ 3   ╳ 4a   ╱ 4b   ▦ 5a   ▥ 5b   ▦ 5c

C ⋯ 6   ▤ 7

D ▤ 8   ▦ 9   ■ 10

0   Km   1000

21 AGRICULTURAL LAND USE AND PRODUCTION

In the USSR as elsewhere, the physical environment plays a major role in determining the broad patterns of land use, though regional and local variations are influenced by economic and other human factors. The predominance of hostile environments over large sections of the country is reflected in the fact that, of the total land area of the USSR (2,228 million hectares), rather less than half (1,048 mill ha) is allocated to farms of all types and less than a quarter (551 mill ha) is used for any agricultural purposes – about a tenth (226 mill ha) is classed as arable and a further 15 per cent (325 mill ha) as meadow, pasture and grazing land. Of the non-agricultural area (1,677 mill ha) about 750 mill ha – one-third of the land surface – is forested, leaving more than 900 mill ha of wasteland. Despite the vast areas of land outside the agricultural sector, agricultural land resources are generous in relation to population numbers; for each Soviet citizen there are just over two hectares of agricultural land, including 0.86 ha of arable, compared with 1.79 ha and 0.83 ha respectively in the United States.

Map 21A indicates regional variations in these broad land use categories. The proportion of agricultural land is highest in the European sector, particularly in the more southerly districts, exceeding 70 per cent in the Black Earth Centre, Ukraine, Moldavia, Volga and North Caucasus regions and varying between 40 and 45 per cent in the Centre, Volga-Vyatka, Urals, Baltic and Belorussia. The arable area is above 40 per cent of the total land area in the first group of regions named and above the national average of 24.5 per cent in the second. As might be expected, the proportion of forest land diminishes southwards from 40–50 per cent in Belorussia, the Volga-Vyatka region and the Centre to 10 per cent or less in the Ukraine and North Caucasus. Wasteland is less than 20 per cent throughout the European sector with the exception of the North-western region. The latter shows a similar situation to that in the three Siberian regions. The North-West, West Siberia, East Siberia and the Far East together constitute some 65 per cent of the land area of the USSR but contain only 13 per cent of its agricultural area and 15 per cent of its arable land. The arable proportion is 7.7 per cent in West Siberia, 2.0 per cent in East Siberia, 1.9 per cent in the North-west and only 0.5 per cent in the Far East region. These regions as a group have more than 40 per cent of their land under forest and there are vast areas completely unused.

Relatively high proportions of agricultural land are recorded in the southerly regions of the USSR – 46 per cent in Transcaucasia, 70 per cent in Kazakhstan and 67 per cent in Central Asia – but these figures are misleading: the majority of the 'agricultural land' is very low grade grazing in arid and semi-arid areas and their arable proportions are low, roughly 13 per cent in each case.

Of the 226 mill ha classed as arable, some 217 mill ha (96 per cent) are sown in any one year, the balance of 9 mill ha remaining fallow. Over the USSR as a whole, 55.9 per cent of the sown area (121.4 mill ha) is devoted to cereals, 6.6 per cent (14.4 mill ha) to industrial crops, 4.0 per cent (8.7 mill ha) to potatoes and other vegetables and 31.0 per cent (67.3 mill ha) to fodders and sown grasses. The dominant trend of recent years has been a decline in the proportion devoted to cereals and a compensating rise in the proportion of the sown area used in the production of feed for livestock, a trend reflected in a rise in the animal protein content of the Soviet food supply. Map 21B indicates regional variations in the uses of the sown area.

Cereals occupy more than half the sown area in the great majority of regions. Proportions below 50 per cent occur under two dissimilar sets of conditions. The area devoted to cereals is relatively low in the more northerly parts of the European USSR (e.g. North-west, Baltic, Belorussia, Centre). These areas, with their cool, damp climates, are better suited to livestock rearing than to food grain production and in recent years have experienced a particularly rapid growth in the area under fodders and sown grasses. They also have large areas under vegetables, especially potatoes.

In more southerly areas – Moldavia, Ukraine, Transcaucasia, Far East and, above all, Central Asia – the low proportion under cereals is due mainly to the importance of the industrial crops. In the extreme case of Central Asia, 43 per cent of the sown area is devoted to such crops, nearly 40 per cent to cotton alone. As a result, this region is deficient in basic foodstuffs, particularly grain, which is 'imported' in large quantities from Kazakhstan.

In Kazakhstan, the Urals, West Siberia and East Siberia, little diversification of agriculture has occurred and cereals are heavily predominant, even by Soviet standards, occupying nearly 70 per cent of the combined sown area of those regions.

The pie-graphs (C) indicate the contributions made to the Soviet agricultural economy by the four main geographical divisions. The European USSR, with barely a quarter of the land area and only about 40 per cent of the agricultural land has nearly 70 per cent of the sown area and contributes about 75 per cent of the total output of crops and livestock products by value. Kazakhstan and Central Asia, despite their large share of the agricultural land and the latter's importance in industrial crop production, contribute only a modest share of total output owing to the predominance of very extensive forms of livestock rearing. The contribution made by Transcaucasia, though small, is much greater than its share of agricultural land or sown area owing to the predominance of high-value sub-tropical crops. Siberia and the Far East, despite their vast size, contribute barely 10 per cent of Soviet agricultural output.

Finally the graphs (D) chart changes in agricultural production during the present century. Until the early 1950s, the increase in output barely kept up with population growth and the food supply available per capita changed very little. Since the mid-1950s, however, the increase in production has been much more rapid, resulting in a more plentiful and more varied diet for the Soviet population. Marked annual fluctuations in the size of the harvest still occur owing to variations in the amount and seasonal distribution of precipitation; these are most clearly visible in the graph of grain output.

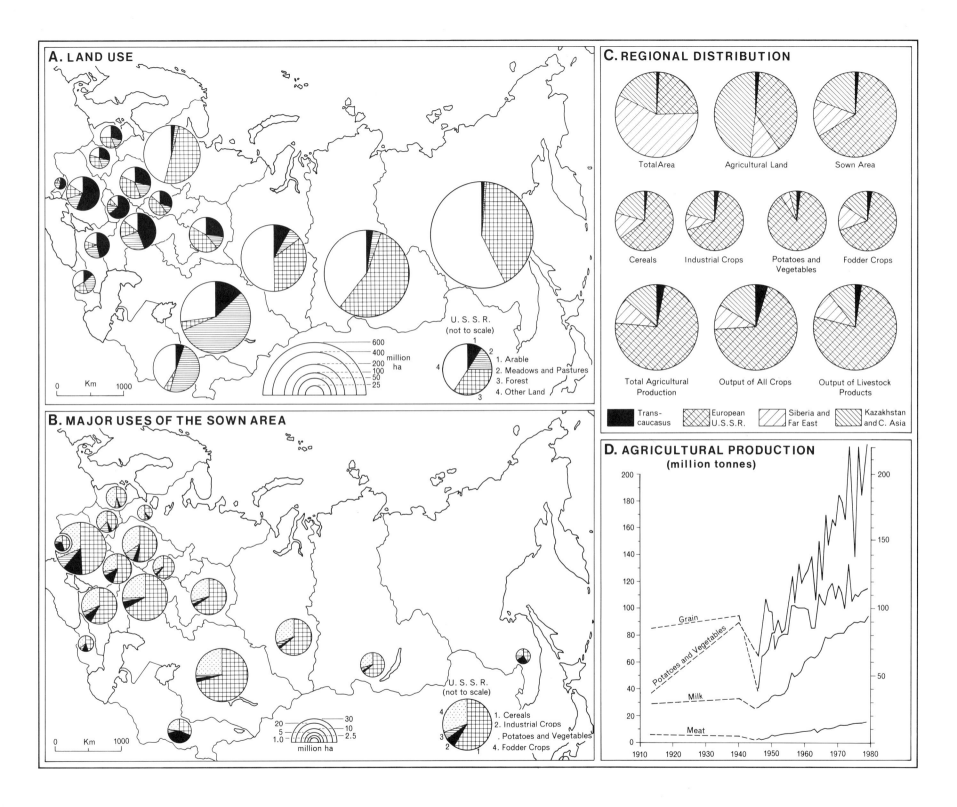

A. LAND USE

U.S.S.R.
(not to scale)

1. Arable
2. Meadows and Pastures
3. Forest
4. Other Land

600
400
200
100
50
25
million ha

0   Km   1000

B. MAJOR USES OF THE SOWN AREA

U.S.S.R.
(not to scale)

1. Cereals
2. Industrial Crops
3. Potatoes and Vegetables
4. Fodder Crops

20   30
5   10
1.0   2.5
million ha

0   Km   1000

C. REGIONAL DISTRIBUTION

Total Area   Agricultural Land   Sown Area

Cereals   Industrial Crops   Potatoes and Vegetables   Fodder Crops

Total Agricultural Production   Output of All Crops   Output of Livestock Products

Trans-caucasus   European U.S.S.R.   Siberia and Far East   Kazakhstan and C. Asia

D. AGRICULTURAL PRODUCTION
(million tonnes)

Grain

Potatoes and Vegetables

Milk

Meat

1910   1920   1930   1940   1950   1960   1970   1980

47

22 MINERALS

The vast size of the country and the great variety of geological structures and rock types within its boundaries ensure that the Soviet Union contains significant deposits of virtually all the minerals required in modern industry. The main deposits currently being worked are shown in Map 22, where the major producing sites for each commodity are named.

Map 22A Iron ore

In view of the high priority given by Soviet economic planners to the development of heavy industry based on large-scale production of steel, it is not surprising that the production of iron ore outweighs all other forms of mineral output. In the period since the Revolution, production of iron ore has multiplied twenty-five times, from barely 10 million tonnes in 1913 to some 250 million tonnes a year in the late 1970s. Over most of this period, the USSR has been able to rely on a small number of sources of high grade iron ore, but since the early 1960s it has been necessary to turn increasingly to lower grade ores; thus in recent years the growth of iron ore production has been more rapid than that of finished metal.

At least half of all Soviet iron ore is mined in the Donets-Dnepr region of the eastern Ukraine, where the Krivoy Rog deposit is the main source and has now been intensively exploited for more than 100 years. Reserves of high grade ore are now being depleted, but there remain vast quantities of low grade iron quartzites which require concentration before they reach the blast furnace.

Until the 1960s, the Urals were the second producer and high-grade magnetite from Magnitogorsk and Nizhniy Tagil supported the bulk of the region's steel production. Output from both these sources has declined and there is increasing reliance on lower grade ores from within the region (e.g. Kachkanar) and from the adjacent Kustanay oblast of Kazakhstan. The Ural region now produces only about 10 per cent of Soviet iron ore compared with 40 per cent in 1950.

A third major source is the Kursk Magnetic Anomaly situated in the Black Earth Centre, midway between Moscow and the Donbass. Although its value was recognised in the 1930s, geological problems and the availability of more easily accessible higher-grade deposits delayed the exploitation of the KMA until the early 1960s, since when it has experienced rapid growth. Output is now in the region of 40 million tonnes, 16 per cent of the Soviet total.

All other sources shown on the map account for less than 20 per cent. Their linkages with steel-producing districts are shown on Map 26 (Iron and steel).

Map 22B Ferro-alloy metal ores

The minerals essential to modern alloy steel production are generally in plentiful supply. There are major sources of manganese at Nikopol in the Ukraine and Chiatura in Georgia and smaller sources in the Urals, Kazakhstan and the Far East. Chrome is mined mainly in the Urals, and nickel in the Urals, the Kola peninsula and (along with copper and platinum) at Norilsk in Arctic Siberia. Molybdenum and tungsten are produced chiefly in the southern mountain zones of Caucasia, Central Asia and Siberia. Several of the rarer items are mined at sites far removed from the main iron and steel producing areas but the high value of these ores outweighs transport costs.

Map 22C Non-ferrous metal ores

Details of non-ferrous metal ore production are absent from most Soviet statistical publications, reflecting the role of these ores as 'strategic materials' and possibly also indicating that some difficulty has been experienced in achieving the desired level of production. Today the USSR is self-sufficient in most items, notable exceptions being bauxite, of which large quantities are imported from Guinea and the Balkans, and tin, imported from Malaysia and Indonesia.

The Ural region is particularly important, with major copper production from several sites as well as platinum, aluminium ores, and cobalt, chrome, titanium and vanadium from the ferro-alloy group. Kazakhstan is a major producer of copper, aluminium ores, lead and zinc, and the Kola peninsula is important for copper as well as for iron and nickel. The Caucasus is another major copper producer. The more northerly parts of East Siberia and the Far East are also rich in minerals but, because of their remoteness, development has so far been confined to the rarer items, notably tin and gold.

Map 22D Non-metallic minerals

Once again the Urals are very important, with large deposits of mineral salts, especially in the Solikamsk-Berezniki district in the north and in the Belaya valley of the Bashkir republic (technically in the Volga region), as well as mica, graphite, asbestos and diamonds.

Mineral salts are also produced in large quantities at Artemovsk in the Donbass, around Lvov in the western Ukraine, in Central Asia and at Usolye Sibirskoye on the Angara river. Sulphur comes mainly from Central Asia and the western Ukraine, while the Yakut republic of the Far East region is noteworthy for the production of diamonds.

It will be observed that, with the exception of iron ore, manganese and some mineral salts, the European plain has limited mineral resources. The richest areas are the Urals and Kazakhstan, but many important items are obtained from more remote areas of the Baltic Shield, the Siberian plateau and the southern mountain zones, districts distant from the industrial areas in which the minerals are used.

As the bar graphs attached to Map 22 show, Soviet production of most items is between 15 and 25 per cent of the world total (a proportion much greater than her share of world population), though bauxite and tin are clear exceptions. The USSR is now the world's leading producer of iron, lead, manganese, mercury, potash and silver and the second or third largest producer of asbestos, chrome, cobalt, copper, diamonds, gold, magnesite, nickel, phosphates, salt, sulphur, tungsten, vanadium and zinc.

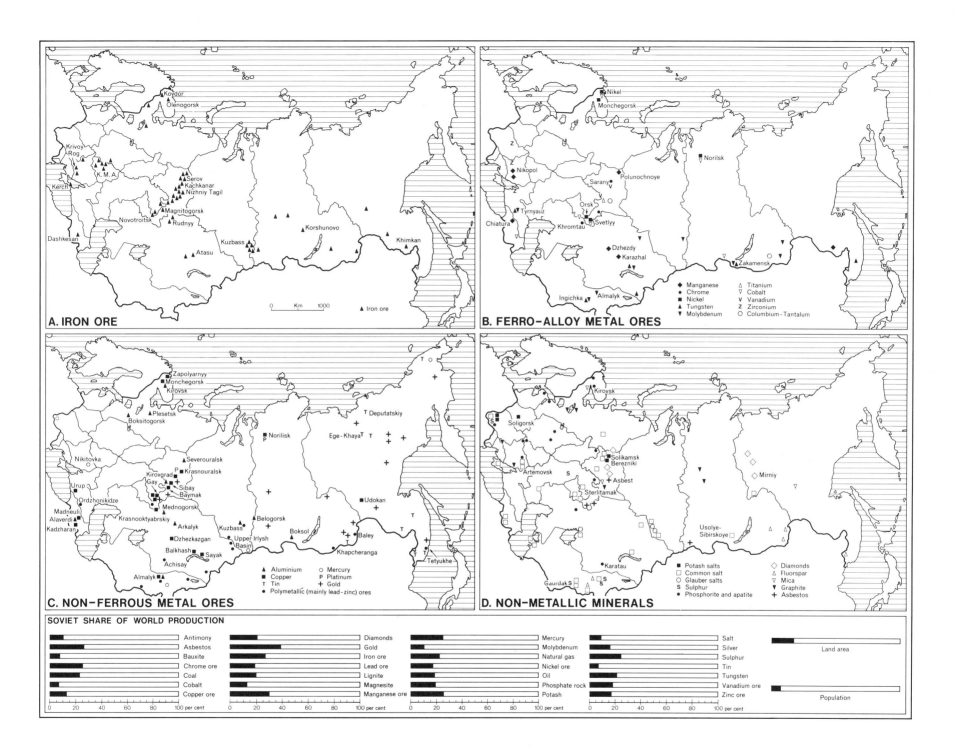

A. IRON ORE

Kovdor
Olenogorsk
Krivoy Rog
K. M. A.
Kerch
Serov
Kachkanar
Nizhniy Tagil
Dashkesan
Magnitogorsk
Novotroitsk
Rudnyy
Korshunovo
Atasu
Kuzbass
Khimkan

0 Km 1000

▲ Iron ore

B. FERRO-ALLOY METAL ORES

Nikel
Monchegorsk
Norilsk
Z
Z
Nikopol
Sarany
Polunochnoye
Orsk
Tyrnyauz
Chiatura
Khromtau
Svetlyy
Dzhezdy
Karazhal
Zakamensk
Ingichka
Almalyk

◆ Manganese	△ Titanium	
● Chrome	▽ Cobalt	
■ Nickel	V Vanadium	
▲ Tungsten	Z Zirconium	
▼ Molybdenum	○ Columbium - Tantalum	

C. NON-FERROUS METAL ORES

Zapolyarnyy
Monchegorsk
Kirovsk
Plesetsk
Boksitogorsk
Deputatskiy
Norilsk
Ege-Khaya
P
Nikitovka
Severouralsk
Kirovgrad
Krasnouralsk
Gay
Sibay
Baymak
Urup
Ordzhonikidze
Madneuli
Mednogorsk
Alaverdi
Krasnooktyabrskiy
Kadzharan
Arkalyk
Kuzbass
Belogorsk
Dzhezkazgan
Upper Irlysh Basin
Boksol
Baley
Balkhash
Sayak
Achisay
Khapcheranga
Almalyk
Tetyukhe

▲ Aluminium	○ Mercury
■ Copper	P Platinum
T Tin	+ Gold
● Polymetallic (mainly lead-zinc) ores	

D. NON-METALLIC MINERALS

Kirovsk
S
Soligorsk
Solikamsk
Berezniki
Artemovsk
S
Asbest
Mirniy
Sterlitamak
Usolye-Sibirskoye
Karatau
Gaurdak S

■ Potash salts	□ Common salt	◇ Diamonds
	○ Glauber salts	△ Fluorspar
S Sulphur		▽ Mica
● Phosphorite and apatite		▼ Graphite
		+ Asbestos

SOVIET SHARE OF WORLD PRODUCTION

Antimony
Asbestos
Bauxite
Chrome ore
Coal
Cobalt
Copper ore

Diamonds
Gold
Iron ore
Lead ore
Lignite
Magnesite
Manganese ore

Mercury
Molybdenum
Natural gas
Nickel ore
Oil
Phosphate rock
Potash

Salt
Silver
Sulphur
Tin
Tungsten
Vanadium ore
Zinc ore

Land area

Population

0 20 40 60 80 100 per cent

49

23 COAL

In the inter-war and early post-war years, coal was the dominant energy source in the USSR. Its share of total energy production rose steadily from 48 per cent in 1913 to a peak of 66 per cent in 1950. Over the same period, annual output increased nearly tenfold, from 29 to 249 million tonnes. Since 1950 there has been a dramatic change resulting from the rapid growth of oil and natural gas production: by 1979, 71 per cent of the country's energy was derived from these sources and the share contributed by coal had fallen to 26 per cent. The decline of coal, however, was only relative: the volume of output has continued to grow and is now well above 700 million tonnes per annum.

The Soviet Union possesses enormous coal reserves – sufficient to support the present level of production for several centuries. A glance at Map 23, however, indicates one major drawback, namely that the great bulk of these reserves lie in the Asiatic regions, whereas the greatest demand is in the more densely settled and highly developed European sector. The coal industry is thus a prime example of one of the great problems facing Soviet economic planners – the vast distances which separate major industrial resources from actual or potential consuming areas. As a result, a situation has developed in which the more conveniently located coalfields, such as the Donbass and Kuzbass, are intensively exploited, while a number of major fields with very large reserves – the Tunguska and Lena basins for example – remain almost untouched owing to their location in remote and environmentally hostile areas.

Coal mining in the Soviet Union today occurs in a large number of widely scattered localities, but output is dominated by half a dozen leading producers. These differ not only in their location and their physical characteristics but also in the nature of their historical development.

Of all Soviet coalfields, the *Donbass* (Donets Basin) of the eastern Ukraine remains the biggest producer, despite a history of intensive exploitation which now covers a period of more than a century. Large-scale mining began here in the 1860s when the region developed as a centre of iron and steel production. For more than 50 years prior to the Revolution, the Donbass was the only major producer, its 1913 output of 25 million tonnes representing 87 per cent of the national total. Since then, while output has risen steadily and now exceeds 200 million tonnes, the Donbass share of the total has fallen below 30 per cent. Production continued to rise, albeit rather slowly, until 1975, despite the physical difficulties presented by seams which are, by Soviet standards, both thin and faulted. The resultant rather high production costs are offset by the high quality of the coals, which include large proportions of coking coal and anthracite, and by the location of the field within the most densely settled and highly industrialised part of the country. In the late 1970s, however, production from the Donbass began to decline.

The *Kuzbass* (Kuznetsk Basin) is by far the most important of the alternative sources of coal developed since the Revolution, its 1979 output of 150 million tonnes representing one-fifth of the Soviet total. Although the Kuzbass was exploited on a small scale in the late nineteenth century, chiefly to supply the Trans-Siberian Railway, development dates mainly from the 1930s when it began to supply coking coal to the iron and steel industries of the Ural region. There was rapid growth during the Second World War when, with the loss of the Donbass, the Kuzbass was the largest producer remaining in Soviet hands, and output has continued to expand in the post-war period. The high proportion of coking coal and the low production costs which result from thick, easily-worked seams offset the transport costs involved in moving the coal to the Urals and elsewhere.

The smaller *Karaganda* field in northern Kazakhstan has a similar history. It was first developed in the 1930s to supply coking coal to

the Magnitogorsk steelworks and continues to fill this role as well as supporting the industrial development of the Kazakh republic.

The *Pechora* field lies in the Arctic north of the European USSR and is noteworthy as an example of large-scale mining in very high latitudes. Developments was mainly during the Second World War under the emergency conditions resulting from the loss of the Donbass. Production costs are high, reserves are limited and little expansion is likely in the near future. However, the technological expertise in high-latitude mining which has accrued from experience in this field will be of great value when and if it becomes necessary to exploit the vast resources of northern Siberia.

The Donbass, Kuzbass, Karaganda and Pechora fields together account for about 60 per cent of Soviet coal production and an even higher proportion of coking coal output. For this reason and because, in addition to supporting local and regional industry, they produce large surpluses for shipment to coal-deficient regions, they are often referred to by Soviet writers as 'coalfields of all-Union importance'.

The remaining areas are of various types. The *Ural* and *Moscow* fields, though ranking high in the list of producers, are of limited value. The coal of the Ural region is generally of low quality, with very little coking coal, and the Moscow field is lignite only, this being used on a large scale in the generation of electricity. Output is declining in both these areas. The *Ekibastuz* and *Kansk-Achinsk* fields, on the other hand, are rapidly expanding their output, despite their low-grade coals and somewhat remote locations. These are examples of a modern trend in the Soviet coal-mining industry, namely the large-scale open-cast mining of low-grade bituminous coal and lignite, which are fed into large thermal electricity generating stations constructed near the pit-head, whence electricity is transmitted to consuming areas through a high voltage grid.

Other coalfields shown on Map 23, contributing in all some 20 per cent of total output, are of local significance only.

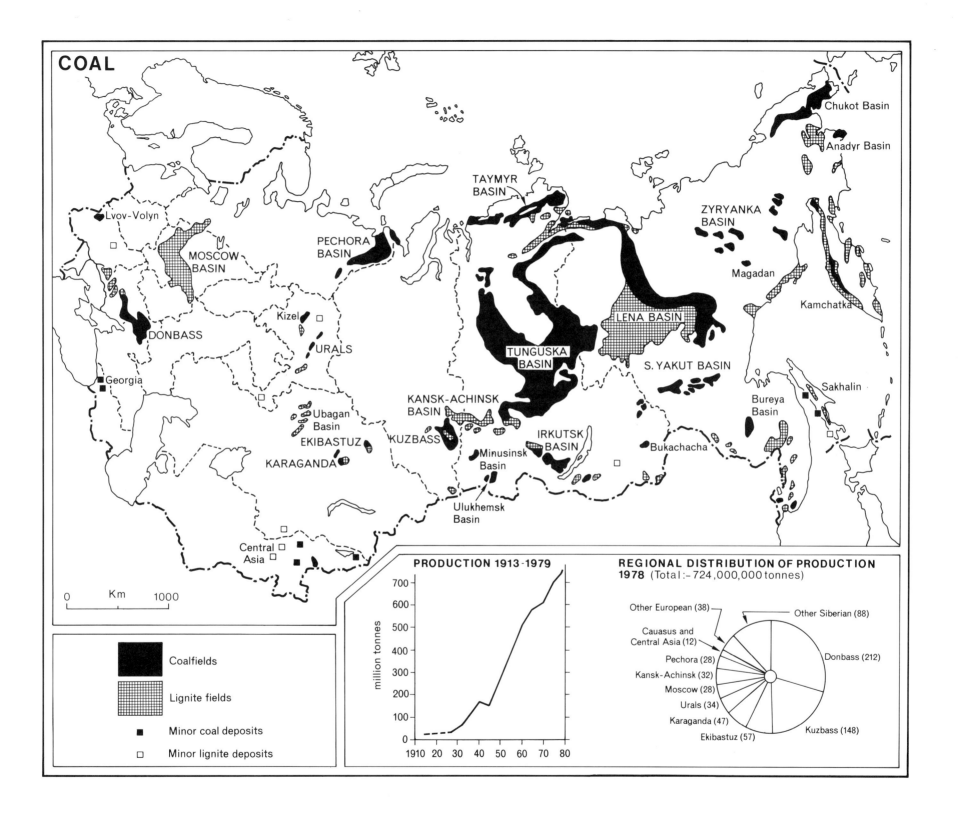

COAL

Chukot Basin

Anadyr Basin

TAYMYR BASIN

ZYRYANKA BASIN

Lvov-Volyn

PECHORA BASIN

Magadan

MOSCOW BASIN

LENA BASIN

Kamchatka

Kizel

DONBASS

URALS

TUNGUSKA BASIN

S. YAKUT BASIN

Georgia

Sakhalin

KANSK-ACHINSK BASIN

Bureya Basin

Ubagan Basin

KUZBASS

Bukachacha

EKIBASTUZ

IRKUTSK BASIN

KARAGANDA

Minusinsk Basin

Ulukhemsk Basin

Central Asia

| 0 | Km | 1000 |

Coalfields

Lignite fields

■ Minor coal deposits

□ Minor lignite deposits

PRODUCTION 1913-1979

million tonnes

700
600
500
400
300
200
100

1910 20 30 40 50 60 70 80

REGIONAL DISTRIBUTION OF PRODUCTION
1978 (Total:– 724,000,000 tonnes)

Other European (38)

Other Siberian (88)

Cauasus and Central Asia (12)

Pechora (28)

Donbass (212)

Kansk-Achinsk (32)

Moscow (28)

Urals (34)

Karaganda (47)

Ekibastuz (57)

Kuzbass (148)

24 OIL AND NATURAL GAS

The text to Map 23 (Coal) has already drawn attention to the rapid increase which has occurred over the last 25 years in the contribution of oil and natural gas to Soviet energy supplies, an aspect further illustrated by the diagram at 24 E. As graphs 24C and 24D clearly show, the production of oil, while of considerable importance, remained at a rather low level prior to 1950 and the output of natural gas was insignificant. Since the latter date, the production of both has grown by leaps and bounds, that of oil rising from 70.8 million tonnes in 1955 to 603 million in 1980 and that of gas from 6,000 to 435,000 million cubic metres. In energy terms, this represents an eleven-fold increase from 112 to 1,250 million tonnes of conventional fuel. Further rapid increases are envisaged during the 1980s. The Soviet Union is now the world's leading oil producer, responsible for nearly 20 per cent of global output (though producing only about half as much as the Middle Eastern countries combined), and second only to the United States for natural gas, contributing nearly a quarter of total world output.

The distribution of Soviet oil and gas fields is indicated in Map 24A, which depicts two major zones. The more southerly stretches from the western flank of the Urals to the middle Volga (Volga-Ural field) and thence down the Volga valley to the Caspian, with extensions westwards to the Black Sea (North Caucasus fields) and eastwards into Central Asia. Immediately east of the Urals the vast Tyumen field underlies most of the West Siberian Lowland. Second-rank fields occur in the Ukraine, the Komi-Ukhta district of north European Russia, in the upper and middle Lena basins of Siberia and on the far eastern island of Sakhalin.

The massive increase of oil and gas production in recent decades has been accompanied by major changes in the relative importance of the various producing districts and in this context it is preferable to consider oil and gas separately.

Oil

Prior to World War II, nearly three-quarters of the Soviet Union's modest oil production came from the Caucasus where the main sites were around Baku, Groznyy and Maykop; in the post-war period, though exploitation of new sources beneath the Caspian and on land has maintained production at 30–40 million tonnes a year, the region's share of total output has fallen to about 5 per cent. In the 1950s and 1960s, the North Caucasus was replaced as the country's leading producer by the Volga-Ural or 'second Baku' field, which by 1960 had an output in excess of 100 million tonnes, some 70 per cent of the Soviet total. This field, too, has now passed its peak and output has been stable at about 200 million tonnes for several years. This now represents about one third of total output and this proportion is declining. The most striking development of the past decade has been the opening up of the vast reserves of West Siberia. Oil was first struck here in 1959 and commercial production began in 1964. Output in 1969 was only 21 million tonnes, but in 1979 it reached 283 million tonnes, nearly half the Soviet total. The remaining fields, though locally important, contribute only 10–15 per cent of the Soviet oil supply.

In the 1980s, the bulk of the increase in output is expected to come from the West Siberian field, with production stable or declining elsewhere. Thereafter the future is uncertain. Reserves in the Lena fields are limited and the Soviet Union may experience difficulty in achieving further increases in output. There may, however, be large offshore resources in the Arctic and the Okhotsk Sea.

Natural Gas

Natural gas shows a similar picture of rapid growth and locational change. In the early stages of development, gas came mainly from established oilfields, and in 1940 more than three-quarters of the small output came from the Baku district. In the early post-war years, the Volga-Ural field became a major gas producer and in the 1950s and early 1960s major discoveries were made in the eastern Ukraine and the Krasnodar and Stavropol districts of the North Caucasus. Since then, the most important developments have been in the northern section of the West Siberian lowland and the deserts of Central Asia; these two areas now produce more than half the Soviet supply and are destined for further increases in output during the 1980s. The most recent major addition to the pattern has been the opening up of a large gas source in the Orenburg oblast of the southern Urals, and there are plans for intensive development in the Yakut (Middle Lena) basin.

Developments in recent decades have resulted in a progressively greater separation between the main oil and gas producing areas and the main consuming centres in the European USSR. This is reflected in the large and expanding pipeline network shown in Map 24B, which connects the oil and gas fields with many of the major cities in the European plain. The system has been extended across the western frontier of the USSR to supply oil and gas to her COMECON partners and to several other European countries. Gas from Iran and Afghanistan is also fed into the Soviet pipeline network.

The growth of oil and gas production has not only added greatly to Soviet energy supplies; it has also been the main support for rapid expansion of the chemical industries. The development of the pipeline system has permitted the establishment of oil refineries, gas-processing plants and petro-chemical complexes in the major industrial areas as well as in oil and gas producing districts.

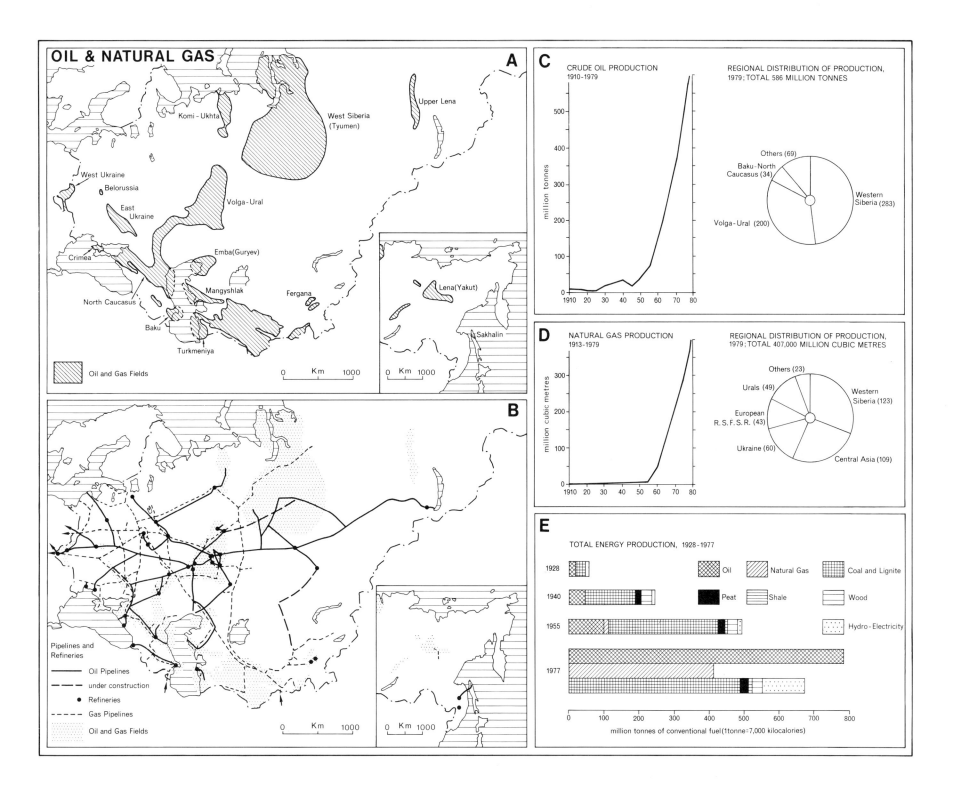

OIL & NATURAL GAS

A

Komi-Ukhta
West Siberia (Tyumen)
Upper Lena
West Ukraine
Belorussia
East Ukraine
Volga-Ural
Crimea
Emba(Guryev)
North Caucasus
Mangyshlak
Fergana
Baku
Turkmeniya
Lena(Yakut)
Sakhalin

Oil and Gas Fields

0   Km   1000

0   Km   1000

B

Pipelines and Refineries

— Oil Pipelines
--- under construction
● Refineries
- - - Gas Pipelines
⋯⋯ Oil and Gas Fields

0   Km   1000

0   Km   1000

C

CRUDE OIL PRODUCTION
1910-1979

REGIONAL DISTRIBUTION OF PRODUCTION, 1979; TOTAL 586 MILLION TONNES

million tonnes

500
400
300
200
100

1910 20 30 40 50 60 70 80

Others (69)
Baku-North Caucasus (34)
Volga-Ural (200)
Western Siberia (283)

D

NATURAL GAS PRODUCTION
1913-1979

REGIONAL DISTRIBUTION OF PRODUCTION, 1979; TOTAL 407,000 MILLION CUBIC METRES

million cubic metres

300
200
100

1910 20 30 40 50 60 70 80

Others (23)
Urals (49)
European R.S.F.S.R. (43)
Ukraine (60)
Western Siberia (123)
Central Asia (109)

E

TOTAL ENERGY PRODUCTION, 1928-1977

1928
1940
1955
1977

▨ Oil   ▨ Natural Gas   ▦ Coal and Lignite
■ Peat   ▤ Shale   ▭ Wood
⋮ Hydro-Electricity

0   100   200   300   400   500   600   700   800
million tonnes of conventional fuel(1tonne=7,000 kilocalories)

25 ELECTRICITY

The development of the electricity supply industry has been a major priority throughout the Soviet period, partly as a result of Lenin's oft-quoted dictum to the effect that 'Communism equals Soviet power plus the electrification of the whole country', and the output of electricity has increased enormously since the Revolution, though from a very low starting point. In 1917, the total installed capacity of all power stations in the USSR was a mere 1.2 million kW (less than that of several individual stations operating today) and output was little more than 2,000 million kWh, only 12.3 kWh per head of the population. By 1980, installed capacity was in excess of 250 million kW and production had reached 1,295,000 million kWh or 4,900 kWh per capita. This last figure was higher than that achieved in such countries as West Germany or the United Kingdom but well behind the 9,400 kWh per capita recorded in the United States.

Soviet electricity is derived from a great variety of energy sources. About 14 per cent comes from hydro-electric stations and five or six per cent from nuclear plant, leaving thermal generating plant responsible for about four-fifths of the total. These use a wide variety of fuels. Despite a rapid growth over the past 25 years in the use of oil and gas for electricity generation, coal remains the most heavily used fuel in thermal plant. There is a large and growing use of low grade fuels. Peat, lignite and low-grade bituminous coal have long been used in large quantities in regions deficient in high grade energy sources such as the Centre and Urals; a more recent development has been the construction of large thermal generating plant at sources of cheap, low-grade fuel outside the main industrial areas, as at Ekibastuz or on the Kansk-Achinsk lignite field, and the transmission of electricity over long distances to consuming areas. Thus Map 25 shows a few major thermal plants in these outlying districts as well as clusters of such plant in the main industrial areas (Centre, Donbass, Urals, Kuzbass) and isolated, generally smaller plant in other urban areas.

The map also indicates the main transmission lines. Prior to the 1950s, transmission took place only over relatively short distances, usually within the confines of a single economic region. During the 1950s and early 1960s, larger regional grids were established for the European USSR, Caucasia, northern Kazakhstan, central Siberia and Central Asia. Links have now been established between these areas and are being strengthened by the construction of very high voltage lines to provide a single transmission system for the whole of the USSR west of Lake Baykal. Soviet electric power is now exported across the western frontier to the COMECON countries of eastern Europe.

Although hydro-electric power supplies less than one-sixth of all electricity generated in the USSR, its exploitation has involved some of the most impressive construction projects in the country. Soviet hydro-electric stations are essentially of two main types. Relatively small stations, often arranged in 'cascades' along a single river, are characteristic of areas where local relief provides a steep gradient, as in Karelia, the Caucasus and the mountains of Central Asia. The really major plant, however, are found mainly in lowland areas, where huge barrages have been constructed across the major rivers to provide an artificial head of water, as on the Dnepr, Volga, Kama, Irtysh, Ob, Yenisey and Angara. The plant at Kuybyshev (Volga), Volgograd, Krasnoyarsk (Yenisey) and Bratsk (Angara) have capacities of 2.3, 2.5, 2.4 and 4.5 million kW respectively, and even larger ones are being built at Ust Ilim (5 million kW) on the Angara and Sayan-Shushenskoye (6.5 million kW) on the upper Yenisey. One or two large plant have also been built or are under construction in the Caucasus, Central Asia and the Far East.

In the cases of the Dnepr (six plant) and the Volga/Kama (eight plant), a succession of barrages has converted the waterway into a string of artificial lakes which have, incidentally, improved navigation and provided irrigation water for the steppe. On the Volga, this has had environmental side effects: extraction of water for irrigation and high evaporation rates from open water surfaces have reduced supplies to the Caspian, presenting problems to navigation and fisheries. As a means of overcoming these difficulties, diversion southward of water from the Pechora and Northern Dvina systems has been suggested.

There are marked regional disparities in the amount of electricity produced per capita. Compared with the national average of 4,900 kWh, certain major industrial areas such as the Centre (4,400) and Ukraine (4,200) are clearly deficient and have to 'import' electricity from other regions. The most outstanding surplus region is East Siberia with a per capita production in excess of 13,500 kWh. Here is another case where the location of resources does not match up with those of population and industry.

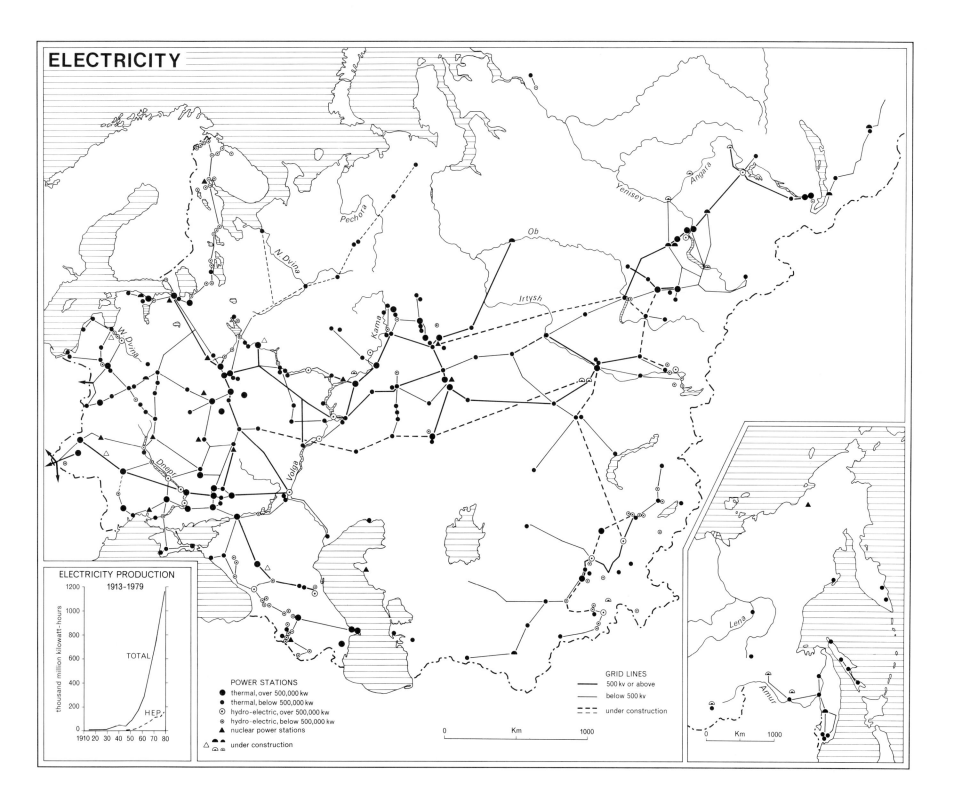

ELECTRICITY

ELECTRICITY PRODUCTION
1913-1979

TOTAL

H.E.P.

thousand million kilowatt-hours

1200
1000
800
600
400
200

1910 20 30 40 50 60 70 80

Pechora

N Dvina

W Dvina

Dnepr

Volga

Kama

Ob

Irtysh

Yenisey

Angara

Lena

Amur

POWER STATIONS
● thermal, over 500,000 kw
• thermal, below 500,000 kw
⊙ hydro-electric, over 500,000 kw
⊙ hydro-electric, below 500,000 kw
▲ nuclear power stations

△ ⌒ ⌒ under construction

GRID LINES
━━━ 500 kv or above
─── below 500 kv
╌╌╌ under construction

0   Km   1000

0   Km   1000

55

26 IRON AND STEEL

Industrial development in the USSR has until recently been primarily concerned with the development of the heavy industrial base. The five-year plans spanning the period from 1928 to the mid-1950s laid particular emphasis on the energy, metallurgical and transport sectors and on heavy engineering. Thus special significance has been attached to the expansion of the iron and steel industry as one of the most basic of industrial activities. Production of steel has been raised from 4.3 million tonnes in 1913 to 148 million tonnes in 1980, making the USSR the world's leading producer. The Soviet iron and steel industry is both widely dispersed and heavily concentrated: activity occurs in a large number of districts – Map 26 shows over 100 sites with one or other branch of the industry – but output is derived mainly from a small number of major producing areas. Two-thirds of Soviet steel comes from the eastern Ukraine and Urals, which are often referred to as the first and second 'metallurgical bases': these titles indicate both their relative importance and order of development.

The *Eastern Ukraine* (the First Metallurgical Base) commenced production in the nineteenth century. Development began in the 1860s, when the coke-smelting process was introduced, and was based on the rich coking coals of the Donbass and the major iron ore deposit at Krivoy Rog in the Dnepr bend. Coal and ore were exchanged by rail and this led to the establishment of steelworks not only on the coalfield but also at Krivoy Rog and several towns along the Dnepr. In the inter-war years, iron ore mining began at Kerch, in the Crimea, and the shipment of iron across the Sea of Azov was followed by the construction of steelworks at Zhdanov on the Ukrainian coast. Additional resources which have been of assistance in the development of the eastern Ukraine as the Soviet Union's leading steel producer have included manganese at Nikopol and Tokmak, iron ore at Kremenchug and Belozerka and hydroelectric power from the Dnepr. Despite the development of other regions, the Ukraine now produces 45 per cent of Soviet pig iron and 37 per cent (56 million tonnes) of the country's steel. As these figures suggest, there is a large movement of pig iron to other areas, notably to steel plant in the Moscow region and in the Soviet Union's COMECON partners.

In contrast, the Second Metallurgical Base in the *Urals* is largely a product of the Soviet period and represents one of the most striking achievements of Soviet economic planning. Iron works were built in the Urals as early as the late seventeenth century, but these were based on charcoal smelting and, in the absence of coking coal, could not adopt the coke-smelting technique. Modern development dates from the 1930s and reflects the post-Revolutionary policy of industrial dispersion. In this case it was considered undesirable that the iron and steel industry should be so heavily concentrated in the Ukraine, a region lying rather close to the strategically vulnerable western frontier. The wisdom of this decision was illustrated during the Second World War, when the Ukraine was occupied by the Germans and the Urals provided most of the steel required by the war effort. The Urals proved extremely rich in iron and various alloy metals but deficient in coking coal. Consequently, the development of the second metallurgical base involved the establishment of links with the Kuzbass and Karaganda, which received Ural iron ore in exchange for coking coal. Today, the movement of coking coal to the Urals continues, but little or no iron ore moves in the opposite direction. In recent years, iron ore production in the Urals has declined with the working out of the more easily accessible high-grade ores at Nizhniy Tagil and Magnitogorsk. Alternative sources have been found within the region, notably at Kachkanar, but there is increasing reliance on outside sources, mainly in northern Kazakhstan. Despite these difficulties, steel output continues to rise, though more slowly than elsewhere, and in 1979 reached 45 million tonnes.

The remaining steel districts are of local or regional rather than national importance, and together contribute only about one-third of total output. Two have already been mentioned: the *Kuzbass* and *Karaganda* produce about ten and five million tonnes per annum respectively. These now rely on local iron ore resources, the Kuzbass also receiving ore from Zheleznogorsk in the Angara valley.

In recent years, production has expanded rapidly in the European RSFSR. The most important district is that centred on *Moscow* (annual output 9–10 million tonnes), where a steel industry has been built up to supply the needs of local manufacturing rather than as a product of local resources, which are in fact very limited. The situation in this part of the country is changing: the last decade has seen the rapid development of the huge ore resources of the KMA (*Kursk Magnetic Anomaly*), situated roughly midway between Moscow and the Donbass, which now supplies one-sixth of Soviet iron ore. As well as supporting local steel plant (e.g. Lipetsk), KMA ores are reaching the Moscow district in increasing quantities. Some Soviet writers believe that future expansion of the steel industry should be based solely on KMA ores, that a halt should be called to the expansion of the industry in the eastern regions and that investment in new steel plant should be confined to the European sector.

There are a number of outlying steel plants which result mainly from the planners' view that each major economic region should have its own steel-producing capacity. The plant at Cherepovets, for example, supplies the north-western region, while Transcaucasia has an iron and steel industry based on local ore and coal. Other 'regional' plants rely on pig iron brought over long distances and scrap metal from local industries and are only small-scale producers: Bekabad in Central Asia, Petrovsk-Zabaykalskiy in eastern Siberia, Komsomolsk-na-Amure in the Far East and Liyepaya on the Baltic coast are in this category.

The current five-year plan (1981–85) envisages no major changes in this pattern: the great bulk of Soviet steel will continue to come from the Ukraine and Urals, supported by much smaller contributions from the Kuzbass, Karaganda, Moscow and KMA districts.

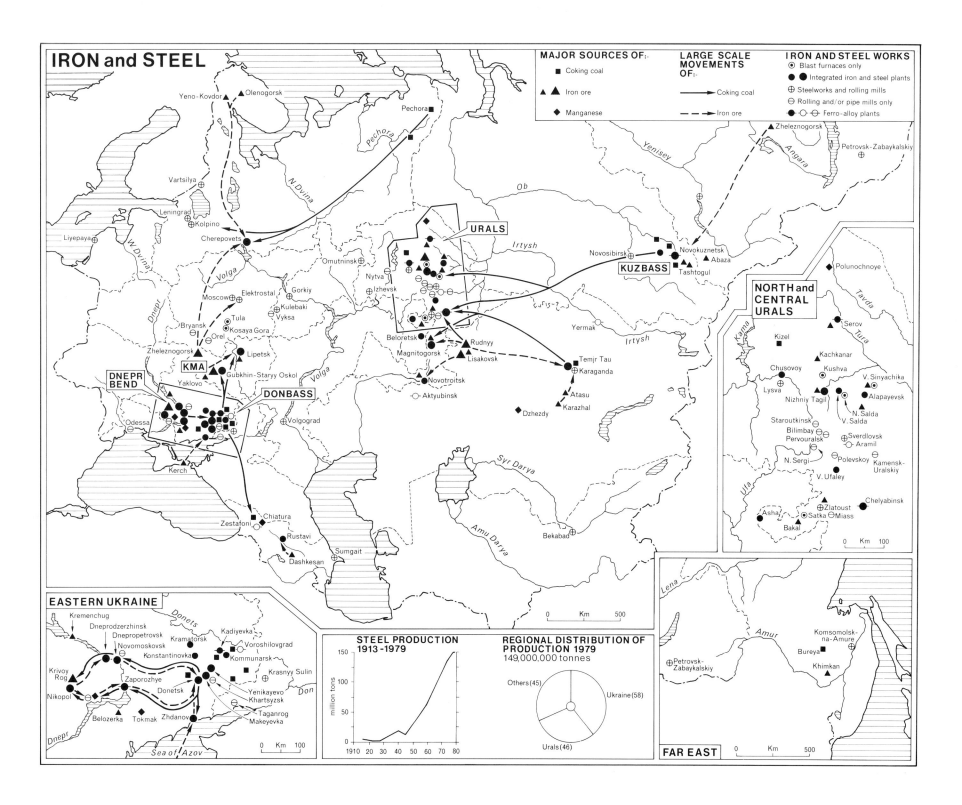

IRON and STEEL

MAJOR SOURCES OF:-
- ■ Coking coal
- ▲ Iron ore
- ◆ Manganese

LARGE SCALE MOVEMENTS OF:-
- ——→ Coking coal
- – – → Iron ore

IRON AND STEEL WORKS
- ⊙ Blast furnaces only
- ● ● Integrated iron and steel plants
- ⊕ Steelworks and rolling mills
- ⊖ Rolling and/or pipe mills only
- ⊕○⊖ Ferro-alloy plants

Yeno-Kovdor ▲   ▲ Olenogorsk
Pechora ■

Vartsilya

Leningrad
⊕ Kolpino
Liyepaya

Cherepovets

W Dvina   N Dvina   Pechora
Volga

Omutninsk ⊕
Moscow ⊕   ⊕ Elektrostal
Gorkiy   Nytva ⊖
⊙ Tula   Kulebaki   Izhevsk
Bryansk   ⊙ Kosaya Gora   Vyksa
Orel

Zheleznogorsk ▲   ● Lipetsk
KMA   Gubkhin-Staryy Oskol
Yaklovo
DNEPR BEND   **DONBASS**

Odessa
Volgograd

Kerch

Chiatura ◆
Zestafoni ⊖   ◆
Rustavi
Dashkesan ▲
Sumgait ⊕

Ob
URALS

Beloretsk
Magnitogorsk ●   ▲ Rudnyy
Lisakovsk
Novotroitsk ●

Yermak ○

Irtysh

Novosibirsk ● ● Novokuznetsk ▲ Abaza
KUZBASS   ▲▲ Tashtogul

Temjr Tau
⊕ Karaganda
▲ Atasu
Karazhal ▲
◆ Dzhezdy

○ Aktyubinsk

Syr Darya

Amu Darya

Bekabad ⊕

▲ Zheleznogorsk
Petrovsk-Zabaykalskiy ⊕
Yenisey   Angara

NORTH and CENTRAL URALS

◆ Polunochnoye

Kama   Tavda   Tura
◆ Kizel
● Serov
▲ Kachkanar
Chusovoy   ⊙ Kushva
Lysva   ▲ V. Sinyachika
Nizhniy Tagil ▲   ▲ Alapayevsk
Staroutkinsk   N. Salda
Bilimbay   V. Salda
Pervouralsk   ⊕ Sverdlovsk
N. Sergi   ○ Aramil
⊖ Polevskoy   ⊖ Kamensk-Uralskiy
● V. Ufaley
▲ Zlatoust   ● Chelyabinsk
Asha ●   ⊙ Satka ⊖ Miass
Bakal ▲

Ufa
0   Km   100

EASTERN UKRAINE

Kremenchug
Dneprodzerzhinsk
Dnepropetrovsk   Donets
Novomoskovsk   Kramatorsk   Kadiyevka
Konstantinovka   ⊕ Voroshilovgrad
Kommunarsk
Krivoy Rog   Krasnyy Sulin ⊕
Zaporozhye   Donetsk
Nikopol ◆   Yenikayevo   Khartsyzsk
Belozerka   Tokmak   Zhdanov   Taganrog   Makeyevka
Dnepr   Don

Sea of Azov
0   Km   100

STEEL PRODUCTION 1913-1979

150	
100	
50	
0	

million tons
1910 20 30 40 50 60 70 80

REGIONAL DISTRIBUTION OF PRODUCTION 1979
149,000,000 tonnes

Others (45)   Ukraine (58)
Urals (46)

FAR EAST

Lena
Amur
Komsomolsk-na-Amure
⊕ Petrovsk-Zabaykalskiy   Bureya ■
Khimkan ▲

0   Km   500

---

0   Km   500

27 NON-FERROUS METALLURGY

As already indicated in our discussion of mineral resources (p. 48), information on the non-ferrous metallurgical sector of Soviet industry is somewhat limited. In comparison with the iron and steel industry, which consumes huge quantities of raw materials and energy to produce 150 million tonnes of metal annually, non-ferrous metallurgy is a relatively small-scale activity, producing in all some 5 or 6 million tons of aluminium, copper, lead, zinc and other metals each year. Nevertheless, these non-ferrous metals are of vital importance to any modern industrial economy; demand for them increases as the economy becomes more sophisticated.

Despite considerable development in the inter-war period, which saw the foundation of all the main branches of non-ferrous metallurgy in the USSR, production of most items remained inadequate for the country's needs and sizeable imports were necessary. Since the mid-1950s there has been accelerated growth in all branches and the Soviet Union is now self-sufficient in most items.

Non-ferrous metal mining and metallurgy as a whole is a widely dispersed activity owing to the variety of ores in use and the several stages of processing involved in the production of finished metal. The earlier stages of treatment, involving the concentration and smelting of ores, usually occur near the raw material source, while refining and metal production may take place near major sources of energy or in the industrial districts which are the markets for non-ferrous metals. In the dispersed pattern of metal production illustrated by Map 27, a number of areas stand out as being of particular importance, notably the Urals, Kazakhstan and southern districts of the West and East Siberian regions. The Kola peninsula, Transcaucasia and Central Asia are also important. A notable feature is the virtual absence of non-ferrous metallurgy from the European regions and the Far East. Thus, apart from the overlap in the Urals, non-ferrous metal-lurgy has a very different location pattern from that of the iron and steel industry.

Copper

In contrast to other branches, copper production in the USSR dates back to the establishment of smelters in the Urals in the seventeenth century. By the eve of World War I, output was more than 30,000 tonnes a year, mostly from the Urals with small-scale production in the Transcaucasus. Expansion in the inter-war period raised production to nearly 150,000 tonnes in 1940. Effort was still concentrated in the Urals but there were also several developments in Kazakhstan. Today, annual output (smelter production) exceeds a million tonnes; the Urals and Kazakhstan remain the chief suppliers, but significant quantities are also produced at Alaverdi in the Caucasus, Almalyk in the Uzbek republic and Norilsk in northern Siberia. There are plans for a major development at Udokan, in the Chita oblast, once the BAM railway (p. 76) is constructed.

Lead and Zinc

Lead and Zinc are derived from 'polymetallic ores', where they occur in conjunction with other metals. Production began on a very small scale in the nineteenth century at Ordzhonikidze in the Caucasus and has increased rapidly in the Soviet period. Output of lead and zinc in the USSR in the late 1970s has been estimated at 620,000 and one million tonnes respectively. The biggest lead smelters are found in the mining districts at Ordzhonikidze, at Chimkent in southern Kazakhstan, at Almalyk and, most important of all, in the upper Irtysh valley in eastern Kazakhstan. Zinc refining, which requires large quantities of electric power is more widely dispersed. The earliest Soviet refineries, using coal and coke distillation processes, were built in the 1930s at Konstantinovka in the Donbass and Belovo in the Kuzbass. The first electrolytic refineries were opened in the same decade at Ordzhonikidze and Chelyabinsk and post-war expansion of zinc production has been assisted by the construction of large electrolytic refineries at Ust-Kamenogorsk, Leninogorsk, Konstantin-ovka and Almalyk.

Aluminium production is the largest and most widespread Soviet non-ferrous metallurgical activity. Output was negligible before World War I and was still only 155,000 tonnes as late as 1950. Since then, production has risen rapidly and exceeded 2.2 million tonnes in 1979. Production involves three stages, the mining of aluminium ores – which, in the USSR, owing to her rather limited bauxite resources, include alternative ores such as nephelite, alunite and kaolinite – the production of alumina and, finally, the conversion of alumina to aluminium, a process requiring large quantities of electric power.

The earliest developments, using the traditional bauxite – alumina – aluminium chain, took place in the 1930s. Bauxite was mined and alumina produced at Boksitogorsk (200 km south-east of Leningrad); the alumina was then converted to aluminium at the Volkhov and Zaporozhye hydro-electric sites. Meanwhile, bauxite had been discovered in the northern Urals and supported an alumina-aluminium plant at Kamensk-Uralskiy. The Volkhov and Zaporozhye plants were destroyed during the World War II but their loss was compensated by the construction of a second Ural alumina-aluminium plant at Krasnoturinsk and an aluminium plant, using alumina from the Urals, at Novokuznetsk in the Donbass.

The massive growth of aluminium production in the post-war period has involved expansion at old sites and the development of new ones. The Boksitogorsk complex now uses apatite from the Kola peninsula and bauxite from Plesetsk in the Onega valley and sends its alumina not only to Volkhov but also to Nadvoitsy and Kandalaksha in Karelia; the Zaporozhye plant, together with a new plant nearing completion at Nikolayev, are based on imported bauxite and that at Volgograd (another HEP site) on alumina from Hungary. Attempts to build up major aluminium-producing complexes in the Transcaucasus and Central Asia based on local nephelite, alunite and kaolinite, however, have been hindered by problems in the development of the appropriate technology to deal with these ores. The most important developments in the post-war period have been the construction of very large aluminium-reduction plants – drawing their alumina from the Urals and Kazakhstan – at the sites of the major hydro-electric projects in southern Siberia. The works at Shelekhov, Bratsk and Krasnoyarsk, all opened in the 1960s, together with the older plant at Novokuznetsk, produce more than half the aluminium output of the USSR.

The developments outlined above have brought the Soviet Union to the position of the world's second producer of the four main items — copper (16 per cent of world production), zinc (15 per cent), aluminium (13 per cent), lead (12 per cent) — of which her output is exceeded only by that of the USA.

NON-FERROUS METALLURGY

Production of:

▲ Aluminium   △ Lead

■ Copper   ● Zinc

□ Nickel   ○ Others

0   Km   1000

28 and 29 THE ENGINEERING INDUSTRIES

The engineering industries as a group constitute the largest single sector of Soviet manufacturing industry and employ about 14 million people — nearly 40 per cent of the industrial labour force and more than 10 per cent of all workers in the USSR. Both the numbers employed and the proportion of the workforce engaged in engineering have risen sharply over the past two decades. In contrast to the consumer goods industries (pages 68–71), where growth was slow prior to the 1950s, the development of the engineering industries has had a high priority and thus a rapid growth throughout the Soviet period.

As the accompanying table shows, the phenomenal growth rates achieved in many branches of engineering prior to 1950 were largely due to the very low level of output before the beginning of the first Five-Year Plan; more significant has been the continuing large-scale absolute growth of production through the 1960s and 1970s.

As the engineering industry has expanded it has also become more diversified and today produces a multitude of different items. Some of these, like machine tools, are used within the engineering industry itself; some, like mining equipment or textile machinery, are the means of production in other industries, both extractive and manufacturing; yet others, like motor vehicles, go direct to the consumer. Thus the markets for engineering products, like the products themselves, are extremely diverse, and the market factor in location operates differently in the various branches. Other locating factors, like labour, raw materials and energy supplies, also vary in their relative importance from one branch to another, and individual engineering activities may be located along a scale from 'heavy', where the amount of energy and raw materials consumed per worker is very large and labour costs are a smaller element, to 'light', where raw material consumption per worker is low and labour costs are dominant. The proportion of skilled labour involved and the level of skills required tend to increase towards the lighter end of the spectrum. In the Soviet case, there has been a marked concentration of effort on the heavier branches, and many of the lighter branches,

particularly those involved in the production of consumer goods, have begun a rapid expansion only during the last two decades. Another factor in the location of the industry has been the Soviet policy of industrial dispersal, and the establishment of engineering activities has often been a major element in the industrialisation of regions formerly lacking modern industry.

For all these reasons, engineering as a whole is a widely dispersed activity and there are few urban centres of any size which do not have at least one industrial enterprise within the broad category of 'engineering'; indeed the establishment of engineering works is commonly the main reason for the classification of a settlement as 'urban'.

Maps 28 and 29 show the distribution of the more important centres engaged in more than a dozen different engineering activities.

Map 28A Heavy engineering

The main activities in this branch of engineering include the production of equipment for the most basic sectors of the economy, namely fuel and power and metallurgy. These types of engineering themselves consume large quantities of energy and metal, especially steel, and thus tend to be concentrated where these are readily available, giving a significant degree of locational concentration. Sites are particularly numerous in, for example, the eastern Ukraine and the Urals.

The production of *metallurgical equipment* is particularly concentrated in these two regions, where the most important centres are Kramatorsk, Kadiyevka, Zhdanov, Dnepropetrovsk, Sverdlovsk and Orsk. Leningrad and Elektrostal (east of Moscow) are the only important sites outside these two regions, though there is also some production in Kazakhstan (Alma-Ata) and Siberia (Usolye Sibirskoye).

The manufacture of *power station equipment* is rather more widely dispersed. The heavier items are made mainly in the chief steel-producing districts, but this branch includes a variety of lighter, more highly skilled products which are made in several less highly industrialised areas.

Mining equipment is produced mainly, though not exclusively, in the major mining areas and

TABLE 7 PRODUCTION IN SELECTED ENGINEERING INDUSTRIES, 1928–1979

	1928	1950	1965	1979
Metal-cutting lathes (000)	2	71	186	280
Forging and pressing machines (000)	0.1	8	35	60
Metallurgical equipment (000 tonnes)	n.d.	111	242	410
Equipment for oil industry (000 tonnes)	1	48	139	188
Equipment for chemical industries (000,000 rubles)	n.d.	n.d.	330	738
Equipment for alimentary industries (000,000 rubles)	n.d.	n.d.	249	523
Equipment for light industries (000,000 rubles)	n.d.	n.d.	281	716
Turbines (000,000 kW)	0.01	2.7	14.6	19.5
Generators and electric motors (000,000 kW)	0.2	7.6	41.3	67.7
Diesel locomotives (000,000 h.p.)	—	0.1	1.5	3.8
Electric locomotives (000,000 h.p.)	—	0.1	0.6	3.5
Railway freight waggons (000)	8	51	40	65
Railway passenger waggons (000)	0.4	0.9	1.8	2.1
Motor cars (000)	0.05	65	201	1,314
Lorries and public service vehicles (000)	0.8	298	415	859
Tractors (000)	1.3	117	355	557
Combines (000)	—	46.3	86.5	115.0
Agricultural machinery (000 million rubles)	n.d.	0.3	1.5	4.7
All engineering (1928 = 1)	1	42	316	1,176

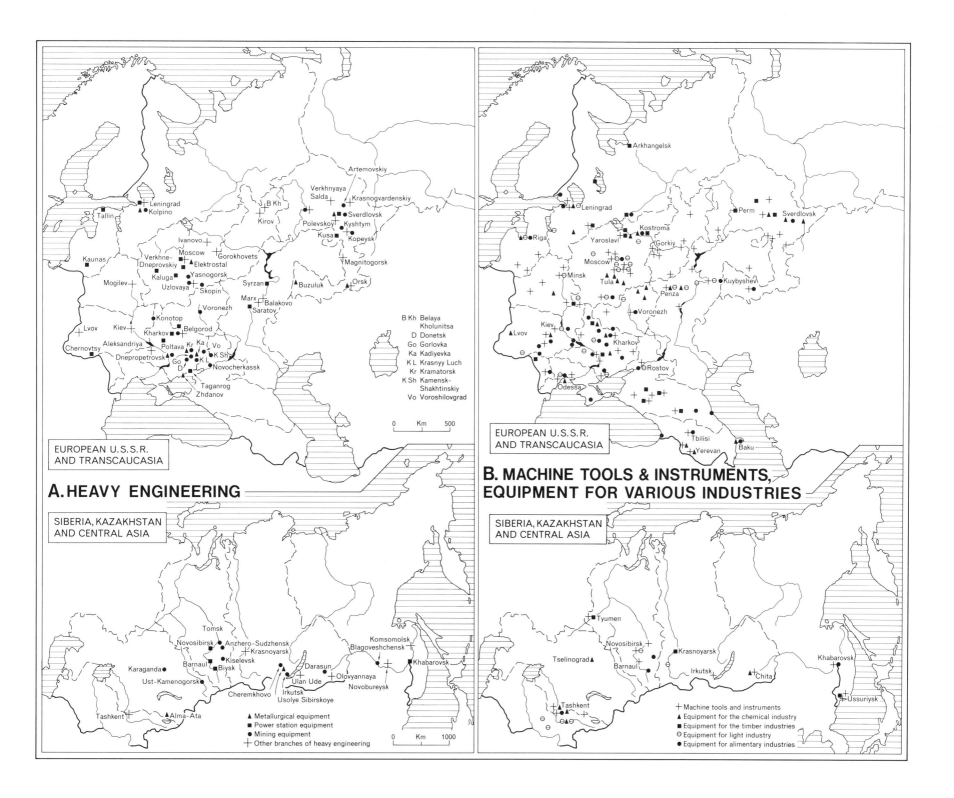

A. HEAVY ENGINEERING

EUROPEAN U.S.S.R.
AND TRANSCAUCASIA

Leningrad
Kolpino
Tallin
Kaunas
Ivanovo
Verkhne-Dneprovskiy
Moscow
Elektrostal
Kaluga
Yasnogorsk
Mogilev
Uzlovaya
Skopin
Lvov
Voronezh
Kiev
Konotop
Kharkov
Belgorod
Chernovtsy
Aleksandriya
Poltava
Kr
Ka
Vo
Dnepropetrovsk
Go
KL
KSh
D
Novocherkassk
Taganrog
Zhdanov
Marx
Balakovo
Saratov
Syrzan
Buzuluk
Orsk
Magnitogorsk
Kusa
Kopeysk
Kyshtym
Polevskoy
Sverdlovsk
Krasnogvardenskiy
Verkhnyaya Salda
Artemovskiy
Kirov
B Kh
Gorokhovets

B Kh Belaya Kholunitsa
D   Donetsk
Go   Gorlovka
Ka   Kadiyevka
K L   Krasnyy Luch
Kr   Kramatorsk
K Sh Kamensk-Shakhtinskiy
Vo   Voroshilovgrad

0 — Km — 500

SIBERIA, KAZAKHSTAN
AND CENTRAL ASIA

Tomsk
Novosibirsk
Anzhero-Sudzhensk
Krasnoyarsk
Kiselevsk
Barnaul
Biysk
Karaganda
Ust-Kamenogorsk
Cheremkhovo
Darasun
Ulan Ude
Irkutsk
Usolye Sibirskoye
Olovyannaya
Novobureysk
Komsomolsk
Blagoveshchensk
Khabarovsk
Tashkent
Alma-Ata

▲ Metallurgical equipment
■ Power station equipment
● Mining equipment
+ Other branches of heavy engineering

0 — Km — 1000

B. MACHINE TOOLS & INSTRUMENTS,
EQUIPMENT FOR VARIOUS INDUSTRIES

EUROPEAN U.S.S.R.
AND TRANSCAUCASIA

Leningrad
Riga
Minsk
Moscow
Yaroslavl
Kostroma
Gorkiy
Tula
Penza
Voronezh
Kiev
Kharkov
Lvov
Odessa
Rostov
Kuybyshev
Perm
Sverdlovsk
Arkhangelsk
Tbilisi
Yerevan
Baku

SIBERIA, KAZAKHSTAN
AND CENTRAL ASIA

Tyumen
Novosibirsk
Barnaul
Tselinograd
Krasnoyarsk
Irkutsk
Chita
Khabarovsk
Ussuriysk
Tashkent

+ Machine tools and instruments
▲ Equipment for the chemical industry
■ Equipment for the timber industries
⊖ Equipment for light industry
● Equipment for alimentary industries

61

there are clusters of sites in the Donbass and Urals as well as individual sites clearly related to local mining activities, for example at Karaganda, Ust-Kamenogorsk and Cheremkhovo.

Map 28B Machine tools and instruments, equipment for various industries

These activities show a very much more widely dispersed location pattern than that of heavy engineering, mainly because of the varied markets to which they supply their products. In view of the large number of sites involved, only the larger centres are named.

The manufacture of *machine tools and instruments*, which are supplied to a wide range of manufacturing industries, including other branches of engineering, is among the most highly skilled engineering activities. Although it occurs widely throughout the European regions, where a large number of new plant have been established over the past 20 years, mainly in medium sized towns, the heaviest concentration is in the Centre, with its tradition of skilled manufacturing industry. There are relatively few such plant in Siberia, Kazakhstan and Central Asia, reflecting the rather selective nature of industrial development of those regions.

The production of equipment for the chemical, timber, light and alimentary industries shows a tendency to concentrate in areas where the manufacturing industries which make use of their products are themselves concentrated. Sites concerned with the manufacture of *equipment for the timber industries* are mainly in the forest zone; the manufacture of *machinery for light industry* – of which textile machinery is a major item – is particularly concentrated in the Centre and, to a lesser degree, Central Asia; equipment for *alimentary industry* is made mainly in southerly areas where the latter is well represented, notably the Ukraine. The production of *equipment for the chemical industries* occurs at numerous sites of different types, reflecting the varied factors which influence the location of the chemical industries themselves (see p. 64).

Map 29A Transport engineering

This is another dispersed activity, each sector of which has its own location pattern.

In the case of *shipbuilding*, the construction of ocean-going vessels is confined to the European regions where the major yards are at Arkhangelsk (White Sea); Leningrad, Tallin, Riga, Kaliningrad (Baltic); Odessa, Nikolayev and Kherson (Black Sea). Vessels for the Caspian are built at Astrakhan and Baku. In addition, there are numerous inland sites on the major rivers.

Railway engineering is also mainly in the European part of the country and is particularly concentrated in the Centre, where diesel locomotives and rolling stock are built in at least 10 towns. Electric locomotives, however, are constructed mainly at Novocherkassk on the Don and at Tbilisi. There are railway repair works in most regions at nodal points on the railway system.

The Soviet *motor vehicle industry* remained very small until the 1960s. The earliest plant were concentrated in the Moscow-Gorkiy zone and additional works were built at Miass in the Urals and Ulyanovsk on the Volga during World War II. In the post-war period, the industry has been established in the Volga region, the Ukraine, Belorussia, Latvia and Georgia, but as yet there are no motor vehicle plant in the eastern regions. Among the largest and most modern works are those at Tolyatti on the Volga, built in collaboration with the Fiat Company, Naberezhniye Chelny on the Kama and Zhodino in Belorussia.

Map 29B Agricultural engineering

This involves the construction of tractors, combines and miscellaneous agricultural equipment. Before World War II there were only four tractor plants – Leningrad, Volgograd, Kharkov and Chelyabinsk – but in the post-war period new works have been opened in the Centre and West Siberian regions and in the Baltic, Belorussian, Moldavian, Ukrainian, Georgian, Kazakh and Uzbek republics. The manufacture of grain combines, which are difficult to transport, is carried on in at least one plant in nearly every region and all regions have centres producing other types of agricultural machinery.

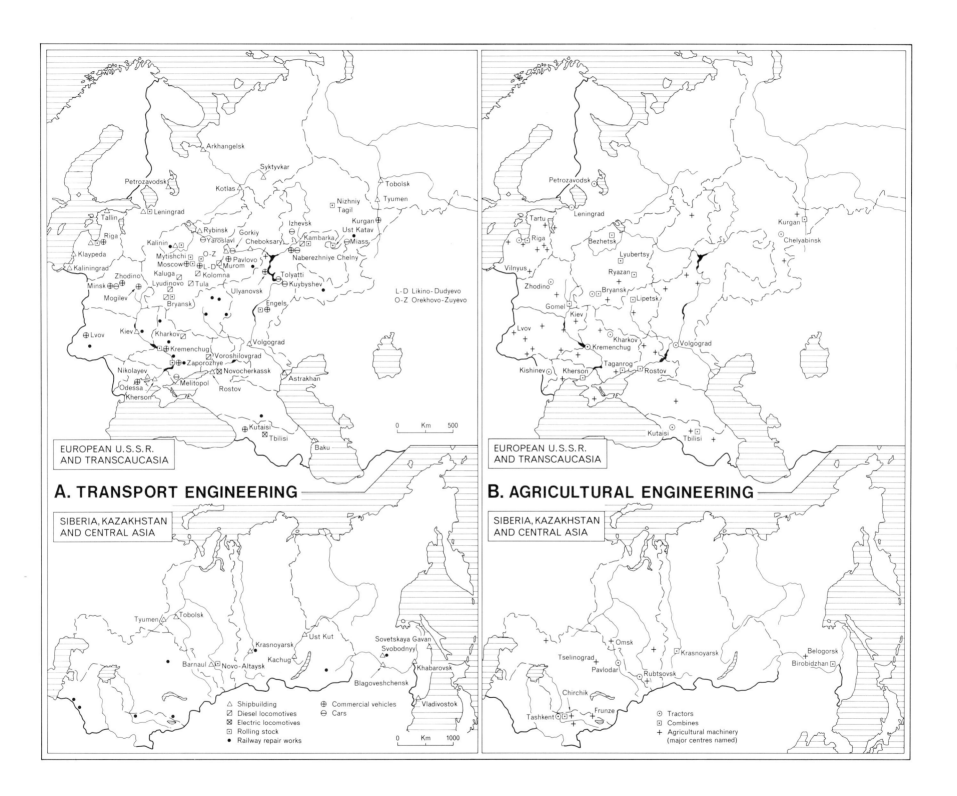

A. TRANSPORT ENGINEERING

EUROPEAN U.S.S.R.
AND TRANSCAUCASIA

SIBERIA, KAZAKHSTAN
AND CENTRAL ASIA

Arkhangelsk
Syktyvkar
Petrozavodsk
Kotlas
Tobolsk
Nizhniy Tagil
Tyumen
Tallin
Leningrad
Kurgan
Ust Katav
Riga
Rybinsk
Gorkiy
Izhevsk
Klaypeda
Kalinin
Yaroslavl
Cheboksary
Kambarka
Miass
Kaliningrad
Mytishchi
O-Z
Pavlovo
Naberezhniye Chelny
Zhodino
Moscow
L-D
Murom
Kaluga
Kolomna
Tolyatti
Minsk
Lyudinovo
Kuybyshev
Mogilev
Tula
Bryansk
Ulyanovsk
Engels
Lvov
Kiev
Kharkov
Volgograd
Kremenchug
Voroshilovgrad
Zaporozhye
Novocherkassk
Nikolayev
Melitopol
Astrakhan
Odessa
Rostov
Kherson
Kutaisi
Tbilisi
Baku

L-D Likino-Dudyevo
O-Z Orekhovo-Zuyevo

Tyumen
Tobolsk
Krasnoyarsk
Ust Kut
Sovetskaya Gavan
Svobodnyy
Barnaul
Novo-Altaysk
Kachug
Khabarovsk
Blagoveshchensk
Vladivostok

△ Shipbuilding
◫ Diesel locomotives
⊠ Electric locomotives
⊡ Rolling stock
● Railway repair works
⊕ Commercial vehicles
⊖ Cars

0   Km   500
0   Km   1000

B. AGRICULTURAL ENGINEERING

EUROPEAN U.S.S.R.
AND TRANSCAUCASIA

SIBERIA, KAZAKHSTAN
AND CENTRAL ASIA

Petrozavodsk
Leningrad
Kurgan
Tartu
Chelyabinsk
Riga
Bezhetsk
Lyubertsy
Vilnyus
Ryazan
Zhodino
Bryansk
Lipetsk
Gomel
Kiev
Lvov
Kharkov
Volgograd
Kremenchug
Taganrog
Kishinev
Kherson
Rostov
Kutaisi
Tbilisi

Omsk
Tselinograd
Krasnoyarsk
Belogorsk
Pavlodar
Birobidzhan
Rubtsovsk
Chirchik
Frunze
Tashkent

⊙ Tractors
⊡ Combines
+ Agricultural machinery
  (major centres named)

63

30 and 31 THE CHEMICAL INDUSTRIES

For the past 25 years, the chemical industries have been among the most rapidly expanding sectors of the Soviet economy. Since 1950, gross output from the chemicals sector has multiplied by a factor of 28, compared with an 11-fold growth for Soviet industry as a whole, and the number employed in chemical manufactures has risen from 350,000 to nearly two million, i.e. from 2.8 to 5.1 per cent of the industrial labour force.

The category 'chemical industries' is an extremely broad one, reflecting the great variety of raw materials used, the many technical processes involved, the wide range of intermediate and final products and the varied markets to which these are supplied. Chemical raw materials include the naturally-occurring minerals shown on Map 22 as well as certain agricultural commodities, notably grain and potatoes. At the same time, the industry both produces and consumes large quantities of chemical compounds which are not readily available in nature but may be derived from minerals, from agricultural commodities or as by-products of oil refining, gas processing, coke making and metal smelting and refining. Thus the oil and gas, coal and coke, iron and steel and non-ferrous metallurgical industries are themselves sources of chemical raw materials. Between the input of raw materials of all types and the output of chemical products there are often several stages of processing, while the final chemical products, though they include many items of direct use to the consumer, are themselves for the most part destined for use in other sectors of industry. Consequently, to indicate a set of location factors applicable to the chemical industries as a whole is virtually impossible and the major branches must be described individually. Before doing so, however, a few general points can be made.

Although chemical industries were established in a few European districts prior to the Revolution and there was considerable expansion in the inter-war years, the Soviet chemical industry was quite small until the 1950s, since when unprecedented growth has occurred. This has been accompanied by rapid technological advance as part of the general modernisation of the Soviet industrial structure during the 1960s and 1970s. Above all, growth has been stimulated and made possible by the tremendous increase in the output of oil and natural gas. While chemical processes based on traditional raw materials are still by no means unimportant, a large and growing proportion are based on oil and natural gas, which have encouraged not only a general increase in the output of chemical products but also diversification into an ever-widening range of commodities. Some idea of the progress achieved is given by the figures in Table 8, which illustrates the particularly rapid growth of the chemical fibres and yarns and the synthetic resins and plastics branches.

TABLE 8 PRODUCTION OF MAJOR CHEMICAL COMMODITIES, 1950–1980 (000 tonnes)

	1950	1965	1980	Increase 1950–80 (times)
Sulphuric acid	2,125	8,518	23,000	10.8
Caustic soda	299	1,199	2,800	9.4
Mineral fertilisers	5,497	31,253	104,000	18.9
Synthetic resins & plastics	67	803	3,600	53.7
Chemical fibres & yarn	24	407	1,200	50.0
Motor tyres (000)	7,400	26,400	60,100	8.1
All chemical products (1950 = 100)	100	766	2,835*	28.4*

* estimated

Map 30A Chemical Industries

Maps 30 and 31 show the distribution of selected major branches of the chemical industry. The first of these (Map 30A) shows all the sites at which chemical industries are indicated in recent Soviet texts and atlases. In view of the great importance of petroleum products, the major oil refineries are also shown (see also Map 24, p. 53). The resultant pattern is a wide dispersal of chemical activities throughout the main settled areas of the USSR. A certain concentration in the vicinity of oil refineries may perhaps be discerned and it should be noted that the latter occur not only on oilfields (several are visible on the Volga-Ural field), but also at other sites (Moscow, Kirishi, Polotsk, etc) on the pipeline network; the use of oil and gas has permitted dispersal as well as expansion of the chemical industries. There are a variety of other situations. The pronounced cluster of chemical sites in the Donbass, for example, reflects the use there of coal/coke by-products and natural mineral salts as well as materials from the Lisichansk refinery; sites in the Moscow basin rely heavily on industrial alcohol from potatoes, local mineral salts and lignite; several of those in the Urals use by-products from non-ferrous metal smelters and refineries; hydro-power sites are also important, as in the Transcaucasus or along the Dnepr.

Map 30B Chemical Fertilisers

The production of *chemical fertilisers* (Map 30B) is by far the biggest of the chemical industries in terms of volume. Prior to the 1960s, low output of mineral fertilisers was a major problem in raising agricultural productivity and this branch has had a high priority. The quantity of mineral fertiliser available per hectare of arable land has risen from 27 kg in 1950 to 490 kg in 1978. There are three main classes of fertiliser, derived respectively from phosphates, potash and nitrogen, each with its own locating factors.

Phosphatic fertilisers are obtained by treating phosphatic raw materials with sulphuric acid. The difficulty of transporting sulphuric acid in bulk and the fact that half the sulphuric acid produced is used in fertiliser production means that these activities are closely related. The traditional source of sulphuric acid in the USSR has been iron pyrites, mined in the Urals and transported to sulphuric acid plants in the main industrial areas. In recent years, other raw materials have become important, notably native sulphur, mined in the Volga valley, at Gaurdak in Central Asia and in the western Ukraine; this is also sent to European sulphuric acid/phosphatic fertiliser plants. The

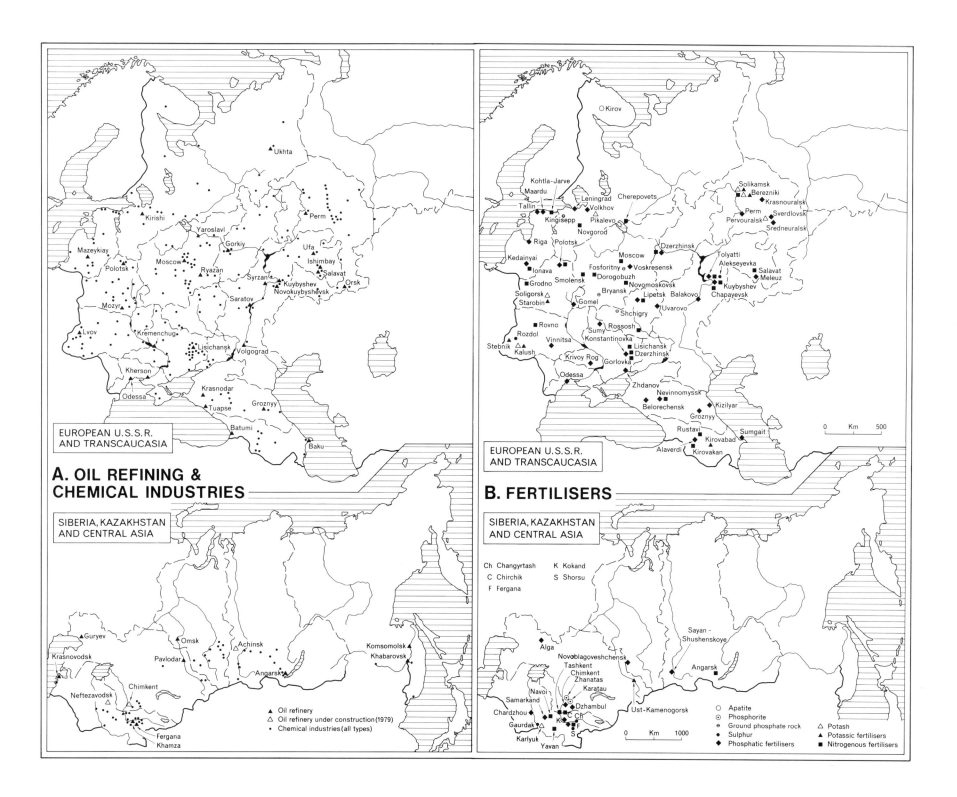

A. OIL REFINING & CHEMICAL INDUSTRIES

EUROPEAN U.S.S.R. AND TRANSCAUCASIA

Ukhta
Kirishi
Perm
Yaroslavl
Gorkiy
Ufa
Mazeykiay
Moscow
Ishimbay
Polotsk
Ryazan
Salavat
Syrzan
Kuybyshev
Mozyr
Novokuybyshevsk
Orsk
Saratov
Lvov
Kremenchug
Lisichansk
Volgograd
Kherson
Krasnodar
Odessa
Grozny
Tuapse
Batumi
Baku

SIBERIA, KAZAKHSTAN AND CENTRAL ASIA

Guryev
Omsk
Achinsk
Krasnovodsk
Pavlodar
Komsomolsk
Khabarovsk
Angarsk
Chimkent
Neftezavodsk
Fergana
Khamza

▲ Oil refinery
△ Oil refinery under construction (1979)
• Chemical industries (all types)

B. FERTILISERS

EUROPEAN U.S.S.R. AND TRANSCAUCASIA

Kirov
Kohtla-Jarve
Maardu
Leningrad
Solikamsk
Berezniki
Tallin
Volkhov
Cherepovets
Krasnouralsk
Kingisepp
Pikalevo
Perm
Pervouralsk
Sverdlovsk
Novgorod
Sredneuralsk
Riga
Polotsk
Dzerzhinsk
Kedainyai
Moscow
Tolyatti
Ionava
Alekseyevka
Fosforitny
Voskresensk
Salavat
Grodno
Smolensk
Dorogobuzh
Meleuz
Soligorsk
Novomoskovsk
Kuybyshev
Starobin
Bryansk
Lipetsk
Balakovo
Chapayevsk
Gomel
Uvarovo
Shchigry
Rovno
Rossosh
Rozdol
Sumy
Stebnik
Vinnitsa
Konstantinovka
Lisichansk
Kalush
Krivoy Rog
Dzerzhinsk
Gorlovka
Odessa
Zhdanov
Nevinnomyssk
Kizilyar
Belorechensk
Grozny
Rustavi
Sumgait
Alaverdi
Kirovabad
Kirovakan

0   Km   500

SIBERIA, KAZAKHSTAN AND CENTRAL ASIA

Ch Changyrtash   K Kokand
C Chirchik   S Shorsu
F Fergana

Sayan - Shushenskoye
Alga
Novoblagoveshchensk
Angarsk
Tashkent
Chimkent
Zhanatas
Karatau
Navoi
Samarkand
Dzhambul
Ust-Kamenogorsk
Chardzhou
Gaurdak
Karlyuk
Yavan

0   Km   1000

○ Apatite
⊙ Phosphorite
⊖ Ground phosphate rock
△ Potash
▲ Potassic fertilisers
◆ Phosphatic fertilisers
■ Nitrogenous fertilisers
● Sulphur

sulphides produced in non-ferrous metal smelting are another source of sulphur used in acid and fertiliser production in the Transcaucasus, Central Asia and the Urals; hydrogen sulphide from oil refining is also important.

Phosphatic fertiliser plant built before the Revolution (e.g. at Leningrad, Riga and Odessa) relied on imported phosphates and there were some inland plant (Vinnitsa, Perm) combining imported phosphates with local phosphate rock. For the past 50 years, however, the main source of phosphates has been apatite concentrate from Kirovsk in the Kola peninsula, which is supplied to fertiliser plant throughout European USSR. New major phosphate sources were developed in the 1960s at Zhanatas and Karatau which supply Central Asian plant.

Potash fertilisers are produced in much smaller quantities at a few sites close to sources of potassium salts. The most important districts are Solikamsk-Berezniki in the Urals and Soligorsk in Belorussia; there is minor production in the western Ukraine, Transcaucasus and eastern Kazakhstan.

The production of *nitrogenous fertilisers*, now the leading branch, involves complex chemical processes. The basic raw material is ammonia, obtained by the combination of atmospheric nitrogen with hydrogen, so the availability of the latter is a major factor. Traditionally, most of the hydrogen used in ammonia synthesis was derived from coal or from coke-oven gases and plant using this process were established in the mining and metallurgical regions of the Ukraine, Centre, Urals, Transcaucasia and Kuzbass. Nowadays, hydrogen is increasingly produced from natural gas. Several of the older plant have been converted to this process and more than a dozen completely new works have been built at sites where gas is available by pipeline: at Kohtla-Iarve (Estonia), Ionava (Lithuania), Grodno (Belorussia), Novgorod, Dorogobuzh (near Smolensk), Rovno, Cherkassy and Lisichansk (Ukraine), Nevinnomyssk (N. Caucasus), Tolyatti, Salavat (Bashkir ASSR) and Fergana, Navoi and Yavan (in Central Asia).

Map 31A Artificial and synthetic fibres

The man-made fibres industry is mainly a product of the post-war period. Prior to World War II, Soviet output was entirely confined to the cellulosic fibre, viscose rayon, produced entirely in the European regions. In the 1950s and 1960s, additional plant were established not only in the European sector but also in the Urals and Siberia and output grew rapidly. As late as 1970, viscose rayon represented 70 per cent of all Soviet man-made fibre production, much of the rest being cuprammonium rayon or acetate rayon. The production of wholly synthetic fibres did not get under way until the late 1960s, since when there has been a massive growth in the output of kapron, lavsan and nitron (the Soviet equivalents of nylon, dacron and orlon). Based entirely on oil and natural gas, plant producing these wholly chemical fibres are widely dispersed, mainly in the European regions.

Map 31B Synthetic rubber, synthetic resins and plastics

Map 31B shows sites engaged in the production of two separate sets of chemical products — synthetic rubber and synthetic resins and plastics. *Synthetic rubber* has been a major element in the Soviet chemical industry since the 1930s, since natural rubber cannot be produced in the USSR for climatic reasons. Several plant were constructed in the European regions producing butadiene-sodium rubber (Buna) based on alcohol obtained from agricultural produce, mainly potatoes. Production of chloroprene (neoprene) rubber using acetylene produced from limestone by the calcium carbide process was established at Yerevan. The post-war period has seen a change-over to rubber produced from oil and natural gas by a variety of processes involving the use of butane, butylene, ethelyne and propylene. Once again, the use of oil and gas has led to a dispersed pattern of production.

The production of *synthetic resins and plastics* has developed mainly since the late 1950s, and these, too, are widely dispersed activities. Urea resins, derived from ammonia, are produced near nitrogenous fertiliser plant, PVC production is associated with the production of chlorine at Novomoskovsk, Yerevan and Usolye Sibiskoye, while plastics based on polyethylene and polypropylene derived from oil refining are made in many areas and constitute the most rapidly expanding sector.

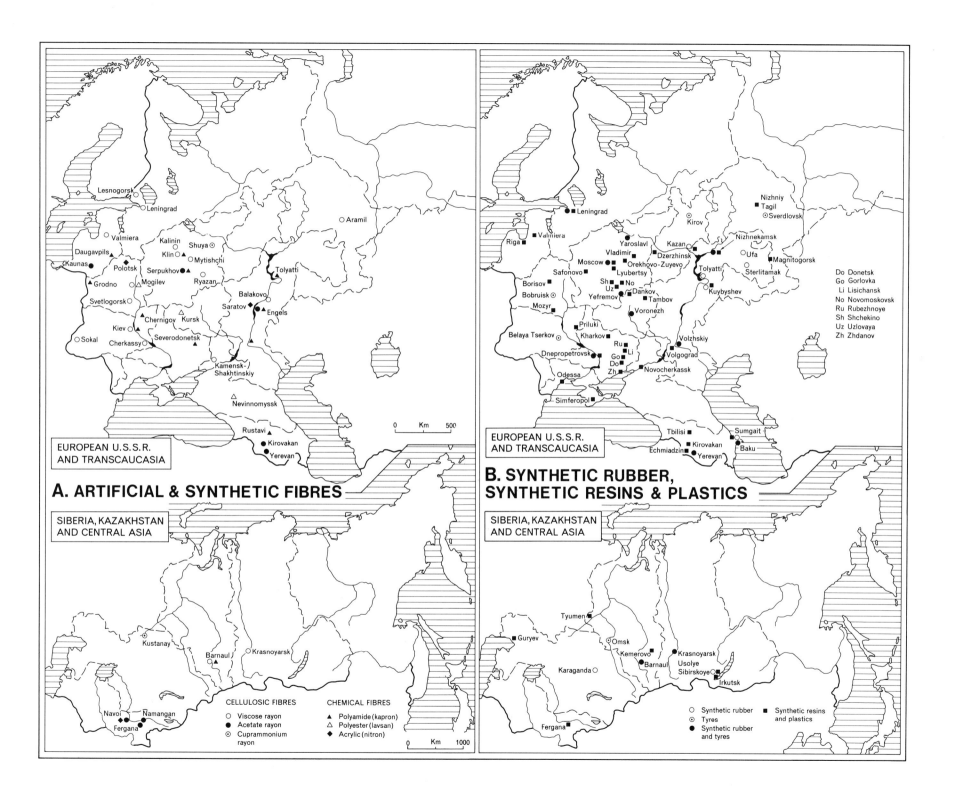

A. ARTIFICIAL & SYNTHETIC FIBRES

EUROPEAN U.S.S.R. AND TRANSCAUCASIA

Lesnogorsk
Leningrad
Valmiera
Daugavpils
Kaunas
Polotsk
Grodno
Mogilev
Svetlogorsk
Sokal
Kiev
Cherkassy
Chernigov
Severodonetsk
Kursk
Kalinin
Klin
Shuya
Serpukhov
Mytishchi
Ryazan
Aramil
Tolyatti
Balakovo
Saratov
Engels
Kamensk-Shakhtinskiy
Nevinnomyssk
Rustavi
Kirovakan
Yerevan

SIBERIA, KAZAKHSTAN AND CENTRAL ASIA

Kustanay
Barnaul
Krasnoyarsk
Navoi
Namangan
Fergana

CELLULOSIC FIBRES
○ Viscose rayon
● Acetate rayon
⊙ Cuprammonium rayon

CHEMICAL FIBRES
▲ Polyamide (kapron)
△ Polyester (lavsan)
◆ Acrylic (nitron)

0 Km 500

B. SYNTHETIC RUBBER, SYNTHETIC RESINS & PLASTICS

EUROPEAN U.S.S.R. AND TRANSCAUCASIA

Leningrad
Riga
Valmiera
Kirov
Nizhniy Tagil
Sverdlovsk
Yaroslavl
Kazan
Nizhnekamsk
Vladimir
Dzerzhinsk
Ufa
Magnitogorsk
Moscow
Orekhovo-Zuyevo
Tolyatti
Safonovo
Lyubertsy
Sh
No
Sterlitamak
Borisov
Uz
Dankov
Kuybyshev
Bobruisk
Yefremov
Tambov
Mozyr
Voronezh
Priluki
Belaya Tserkov
Kharkov
Ru
Li
Volzhskiy
Dnepropetrovsk
Go
Do
Zh
Volgograd
Novocherkassk
Odessa
Simferopol
Tbilisi
Sumgait
Kirovakan
Echmiadzin
Yerevan
Baku

Do Donetsk
Go Gorlovka
Li Lisichansk
No Novomoskovsk
Ru Rubezhnoye
Sh Shchekino
Uz Uzlovaya
Zh Zhdanov

SIBERIA, KAZAKHSTAN AND CENTRAL ASIA

Tyumen
Guryev
Omsk
Kemerovo
Krasnoyarsk
Barnaul
Usolye Sibirskoye
Karaganda
Irkutsk
Fergana

○ Synthetic rubber
⊙ Tyres
● Synthetic rubber and tyres
■ Synthetic resins and plastics

0 Km 1000

32 and 33 TEXTILE, CLOTHING AND FOOT-WEAR INDUSTRIES

In the USSR since the Revolution, the production of consumer goods ('Group B industries') has expanded much less rapidly than that of energy, raw materials and capital equipment ('Group A industries'), though in recent years the gap in growth rates between these two has perceptibly narrowed. Between 1928 – the beginning of the first Five Year Plan – and 1978, gross output of all Soviet industry multiplied by a factor of 131, including a 274-fold increase in production by Group A industries and a 42-fold increase in Group B, a ratio of 6.5 to 1 in favour of Group A. Since 1960, total output has multiplied 3.7-fold, 4.1 times for Group A and 3.2-fold in Group B, a ratio of only 1.3 to 1. Though Group A industries still represent by far the greater part of the Soviet Union's industrial capacity and production and receive well over 80 per cent of all capital investment in industry, the accelerated growth of Group B industries is now producing a more plentiful supply of consumer goods for the Soviet population.

Of all the consumer goods industries, the manufacture of textiles is among the few for which detailed data are regularly published in Soviet statistical handbooks, and textiles, clothing and footwear are the only items appearing on maps depicting the distribution of 'light industry' in Soviet texts and atlases. Textile production trends since the Revolution (Table 9) are characteristic of the consumer goods sector as a whole, showing relatively slow growth until the 1950s, followed by rapid expansion over the past 25 years. From 1928 to 1950, textile production rose by only 54 per cent; since 1950 output has trebled. Maps 32 and 33 indicate the location of centres engaged in the production of various types of textiles, clothing and footwear.

Map 32A Cotton textiles

As the accompanying table shows, cotton fabrics

TABLE 9 TEXTILE PRODUCTION, 1928–1979
(millions of square metres)

	1928	1950	1965	1979
Cotton	1,820	2,745	5,499	6,977
Woollen	112	193	466	774
Linen	177	257	548	768
Silk and artificials	8	106	801	1,724
Hemp and jute	80	73	158	147
All Types	2,197	3,374	7,472	10,390

represent about two-thirds of all textile production in the USSR, though this proportion has declined considerably in recent years with the rapid expansion of the artificials. A mechanised cotton textile industry was established in the nineteenth century in the European Centre – a region with a long tradition of domestic wool and linen manufacture – drawing its raw cotton from the newly-acquired territories of Central Asia and Transcaucasia. In the Soviet period, despite the opening of large mills in the cotton-growing areas, production of cotton cloth has remained heavily concentrated in the European regions, which are responsible for 83 per cent of total output, 70 per cent from the Centre alone. The Central Asian republics and Transcaucasia produce only 12 per cent, the balance coming from a few sites in Siberia. Map 32A shows a particularly dense concentration in a zone between the upper Volga and Oka rivers where there are nearly 30 cotton textile towns, of which Ivanovo, Kostroma and Yaroslavl are the biggest. This sector of Soviet industry displays a remarkable degree of geographical inertia in the persistence of a location pattern established more than a century ago.

Map 32B Other textiles

Map 32B shows the distribution of centres producing other types of textiles – woollen, linen, silk and artificials, hemp and jute.

Woollen textile manufacture is even more heavily concentrated in the European regions than that of cotton cloth. Some 50 per cent of

total output comes from the Centre, where there is a concentration of sites immediately to the east of Moscow, and another 40 per cent from scattered centres in Belorussia, the Ukraine, the North-west, Urals, Volga and North Caucasus. There are also a few producing centres in Siberia, Central Asia and Transcaucasia, but these Asiatic regions together account for barely 10 per cent of total output.

Linen textile manufacture is also found mainly in the forest zone of the European USSR where flax was the main industrial crop in the traditional agricultural system and a domestic linen industry was established at an early stage. Two-thirds of total output comes from the Centre, where there is a heavy concentration of activity in the eastern part of the Volga-Oka zone, and a further 20 per cent from flax-growing districts of the North-west, Baltic and Belorussia regions. In marked contrast to cotton manufacture, the linen branch is found close to its raw material supply.

Soviet statistical sources often combine *silk and artificial* textiles, but the bulk of the rapid growth that has occurred in this sector in recent years can be attributed to the increased output of the artificials. In the latter case, there is an overlap with the chemical industries and the location of plants producing artificial and synthetic fibres (as distinct from cloth) is shown on Map 31A, p. 67. Much of the artificial and synthetic fibre produced goes to the manufacture of cloth in which these fibres are combined with natural ones, but a small number of large, modern plant produce wholly artificial cloth. These are in the Centre, which accounts for 60 per cent of output in the silk and artificials sector; a large plant also exists at Krasnoyarsk in Siberia.

Natural silk is produced, on a much smaller scale, at numerous, generally small, plant in Transcaucasia and Central Asia where silk is a traditional product.

Finally, *hemp and jute* textiles are produced at a small number of centres – hemp at Saransk (Mordov ASSR) and Frunze (Kirgiz republic), both in hemp-growing districts; and jute at Odessa, where the imported raw material is landed.

Thus the textile industry as a whole is very

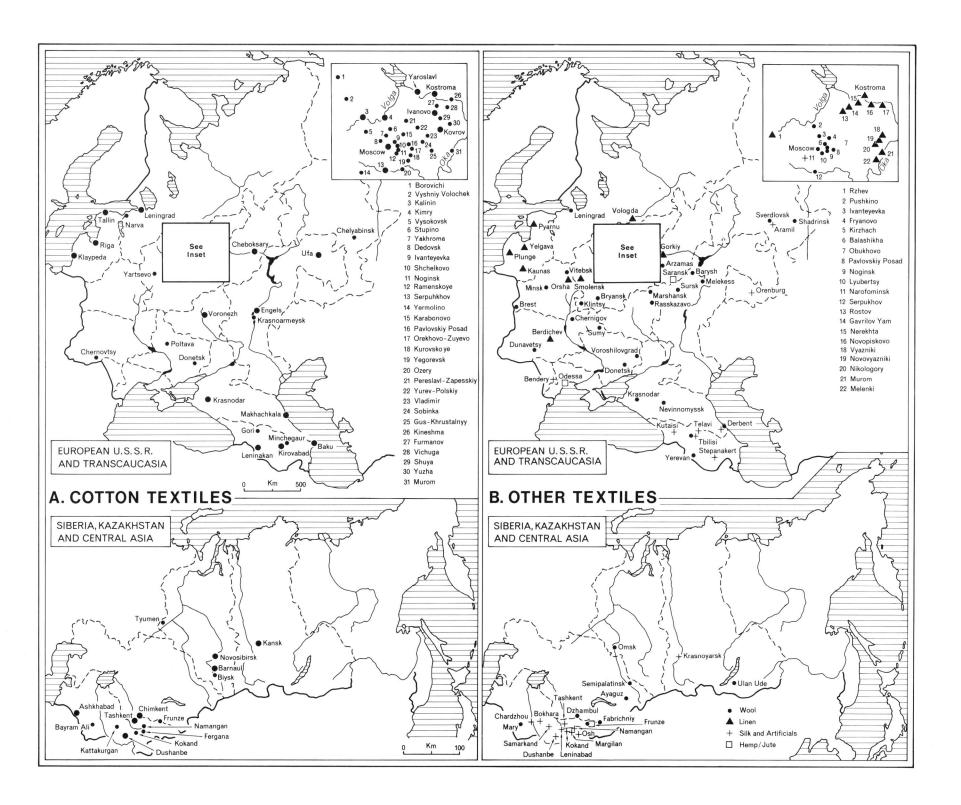

A. COTTON TEXTILES

EUROPEAN U.S.S.R. AND TRANSCAUCASIA

1 Borovichi
2 Vyshniy Volochek
3 Kalinin
4 Kimry
5 Vysokovsk
6 Stupino
7 Yakhroma
8 Dedovsk
9 Ivanteyevka
10 Shchelkovo
11 Noginsk
12 Ramenskoye
13 Serpukhov
14 Yermolino
15 Karabonovo
16 Pavlovskiy Posad
17 Orekhovo-Zuyevo
18 Kurovskoye
19 Yegorevsk
20 Ozery
21 Pereslavl-Zapesskiy
22 Yurev-Polskiy
23 Vladimir
24 Sobinka
25 Gus-Khrustalnyy
26 Kineshma
27 Furmanov
28 Vichuga
29 Shuya
30 Yuzha
31 Murom

SIBERIA, KAZAKHSTAN AND CENTRAL ASIA

B. OTHER TEXTILES

EUROPEAN U.S.S.R. AND TRANSCAUCASIA

1 Rzhev
2 Pushkino
3 Ivanteyevka
4 Fryanovo
5 Kirzhach
6 Balashikha
7 Obukhovo
8 Pavlovskiy Posad
9 Noginsk
10 Lyubertsy
11 Narofominsk
12 Serpukhov
13 Rostov
14 Gavrilov Yam
15 Nerekhta
16 Novopiskovo
18 Vyazniki
19 Novovyazniki
20 Nikologory
21 Murom
22 Melenki

SIBERIA, KAZAKHSTAN AND CENTRAL ASIA

● Wool
▲ Linen
+ Silk and Artificials
□ Hemp/Jute

69

heavily concentrated in the old-established European industrial areas. 70 per cent of all textiles are produced in the Centre, Volga-Vyatka and Central Black Earth regions, about 18 per cent in other European regions, less than 10 per cent in Transcaucasia and Central Asia and barely 3 per cent in Siberia and the Far East.

Maps 33A and 33B Clothing and Footwear

Production trends have been similar to those in the textile industries. The rise in production down to the 1950s barely kept pace with population growth but during the 1960s and 1970s output of most items has trebled and the per capita supply has at least doubled. In contrast to the textile industries which, as we have seen, are heavily concentrated in a few producing districts, these activities are widely dispersed, occurring for the most part near their markets with plant in the majority of large and medium-sized towns throughout the USSR.

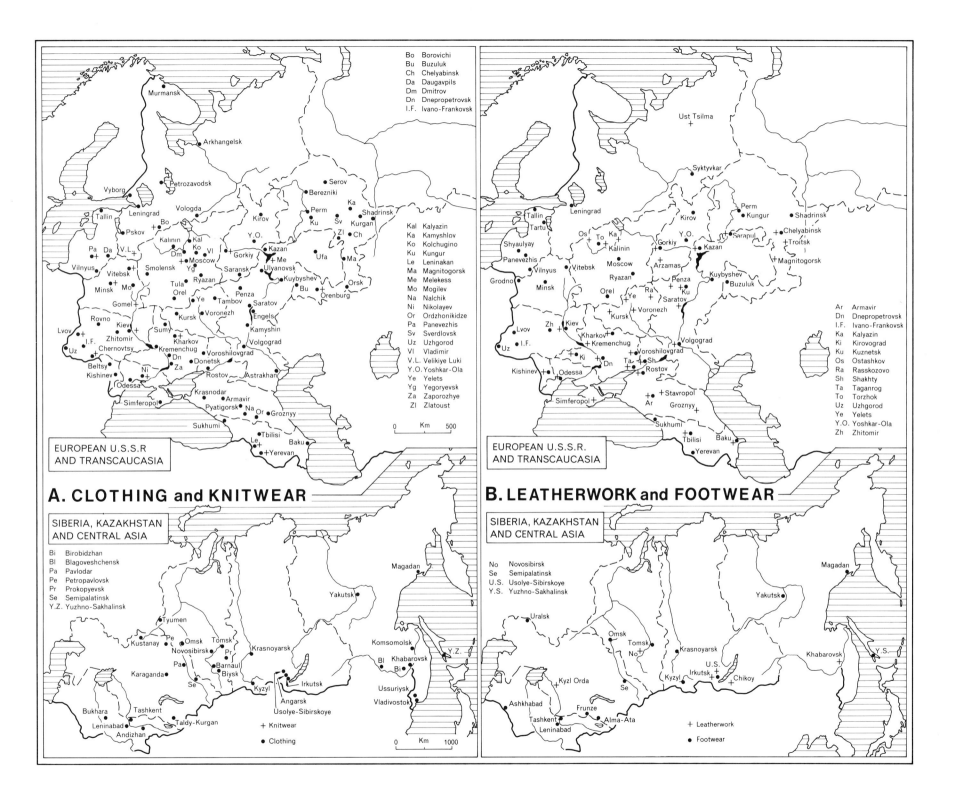

Bo Borovichi
Bu Buzuluk
Ch Chelyabinsk
Da Daugavpils
Dm Dmitrov
Dn Dnepropetrovsk
I.F. Ivano-Frankovsk

Kal Kalyazin
Ka Kamyshlov
Ko Kolchugino
Ku Kungur
Le Leninakan
Ma Magnitogorsk
Me Melekess
Mo Mogilev
Na Nalchik
Ni Nikolayev
Or Ordzhonikidze
Pa Panevezhis
Sv Sverdlovsk
Uz Uzhgorod
Vl Vladimir
V.L. Velikiye Luki
Y.O. Yoshkar-Ola
Ye Yelets
Yg Yegoryevsk
Za Zaporozhye
Zl Zlatoust

Ar Armavir
Dn Dnepropetrovsk
I.F. Ivano-Frankovsk
Ka Kalyazin
Ki Kirovograd
Ku Kuznetsk
Os Ostashkov
Ra Rasskozovo
Sh Shakhty
Ta Taganrog
To Torzhok
Uz Uzhgorod
Ye Yelets
Y.O. Yoshkar-Ola
Zh Zhitomir

EUROPEAN U.S.S.R. AND TRANSCAUCASIA

EUROPEAN U.S.S.R. AND TRANSCAUCASIA

A. CLOTHING and KNITWEAR

B. LEATHERWORK and FOOTWEAR

SIBERIA, KAZAKHSTAN AND CENTRAL ASIA

SIBERIA, KAZAKHSTAN AND CENTRAL ASIA

Bi Birobidzhan
Bl Blagoveshchensk
Pa Pavlodar
Pe Petropavlovsk
Pr Prokopyevsk
Se Semipalatinsk
Y.Z. Yuzhno-Sakhalinsk

No Novosibirsk
Se Semipalatinsk
U.S. Usolye-Sibirskoye
Y.S. Yuzhno-Sakhalinsk

+ Knitwear
• Clothing

+ Leatherwork
• Footwear

0   Km   500

0   Km   1000

71

34　TIMBER INDUSTRIES

As might be assumed from our earlier discussions of vegetation (p. 16) and land use (p. 46), timber is a major natural resource of the USSR, where approximately one-third of the land surface — more than 700 million hectares — carries forest of one kind or another. As Map 34A clearly indicates, land under forest exceeds 20 per cent of the total area throughout virtually the whole of the Russian, Belorussian and Baltic republics, and exceeds 40 per cent in the North-eastern part of the European USSR and over most of East Siberia and the Far East. In contrast, less than 10 per cent is forested in the Ukraine and less than one per cent in Kazakhstan and Central Asia and in the tundra zone of the far north.

Soviet timber production (Table 10) rose steadily until the mid-1960s since when it has remained fairly stable at 350–380 million cubic

TABLE 10 OUTPUT OF TIMBER PRODUCTS, 1928–1979

	1928	1950	1965	1979
Raw timber (mill cu m)	62	266	379	354
Commercial timber (mill cu m)	36	161	274	273
Firewood (mill cu m)	26	105	105	78
Sawn timber (mill cu m)	14	50	110	100
Cellulose (mill tonnes)	0.1	1.1	3.2	7.0
Paper & Cardboard (mill tonnes)	0.3	1.5	4.7	8.7

metres a year, about 15 per cent of the world total. By far the greater part of this output — some 325 mill cu m — is softwood derived from coniferous trees, of which the USSR is the world's leading producer, accounting for 30 per cent of world output. Soviet hardwood production is much smaller — about 65 mill cu m or 4.5 per cent of the world total. Although only four or five per cent of the timber produced is exported, this represents about one-fifth of world trade in softwoods and the USSR is second in this respect only to Canada.

Recent decades have seen a pronounced change in the location of Soviet timber production and an increasing imbalance between producing and consuming areas. Siberia and the Far East contain some 80 per cent of the country's timber reserves but still account for only about 35 per cent of output. Two-thirds of total production and three-quarters of consumption are in the European regions, where the annual cut now exceeds the rate of natural regeneration, and re-afforestation schemes have been undertaken in some of the most heavily exploited areas. In Siberia, on the other hand, the annual cut is barely 10 per cent of annual growth. Consequently, the lumbering industry has moved to progressively more remote parts of the European tayga and, increasingly, to Siberia, resulting in a rapid increase in the cost of moving timber to the markets. Of the total timber felled, about 80 per cent is now used commercially; the proportion used as firewood has declined from 40 per cent in 1950 to 20 per cent in the late 1970s.

Map 34B shows the location of lumbering and saw-milling activities. *Lumbering* in the European regions is now almost entirely north of 56° N, where the most important areas are in the Arkhangelsk, Vologda, Kirov, Perm and Sverdlovsk oblasts and the Karelian and Komi ASSRs. In Siberia, development has been most intensive in the more southerly parts of the tayga, notably in the Tyumen, Tomsk, Krasnoyarsk and Irkutsk areas. Raw timber is moved by rail and water from lumbering sites to a smaller number of *saw-milling* centres. These are mainly in the forest zone, especially near sources of energy, but outlying centres occur much further south, for instance along the Volga river. *Woodworking* industries (Map 34C) on the other hand, are essentially market-oriented and are thus widely dispersed throughout the main settled zones.

The *pulp and paper* industries (Map 34D), which have expanded rapidly during the 1970s (Table 10), consume large quantities of timber, power and water, and are therefore less scattered than other timber industries. About a third of total paper production occurs in the North-western region and another 20 per cent in the Urals. A recent development has been the construction of major pulp mills in Siberia. Finally, *timber-based chemical industries* are found mainly at sites in the forest zone, where they are often associated with other chemical activities as in the North-west, Urals, Volga-Vyatka and Belorussia and close to major power sites in Siberia, as well as in chemical districts of the Volga region and elsewhere.

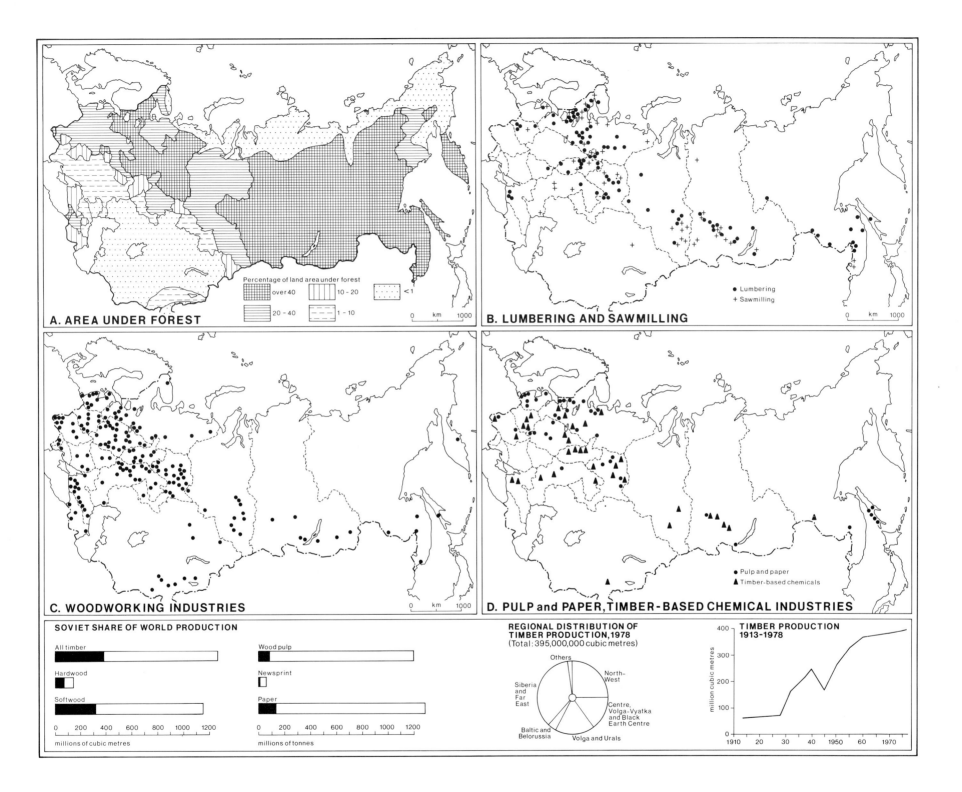

A. AREA UNDER FOREST

Percentage of land area under forest
over 40
20 - 40
10 - 20
1 - 10
< 1

0 km 1000

B. LUMBERING AND SAWMILLING

● Lumbering
+ Sawmilling

0 km 1000

C. WOODWORKING INDUSTRIES

0 km 1000

D. PULP and PAPER, TIMBER-BASED CHEMICAL INDUSTRIES

● Pulp and paper
▲ Timber-based chemicals

SOVIET SHARE OF WORLD PRODUCTION

All timber

Hardwood

Softwood

0 200 400 600 800 1000 1200
millions of cubic metres

Wood pulp

Newsprint

Paper

0 200 400 600 800 1000 1200
millions of tonnes

REGIONAL DISTRIBUTION OF
TIMBER PRODUCTION, 1978
(Total: 395,000,000 cubic metres)

Others
North-West
Centre, Volga-Vyatka and Black Earth Centre
Volga and Urals
Baltic and Belorussia
Siberia and Far East

TIMBER PRODUCTION
1913-1978

400
300
200
100
0

million cubic metres

1910 20 30 40 1950 60 1970

73

35 ALIMENTARY INDUSTRIES

Production trends in this sector (Table 11) are similar to those recorded by other consumer goods industries. Output of most items was at a very low level in the 1920s and had barely doubled by 1950, indicating only a slow increase in per capita consumption of processed foodstuffs. Over the past 25 years, progress has been much more rapid; gross output of the alimentary industries as a whole has multiplied five-fold since 1955.

TABLE 11 PRODUCTION IN MAJOR BRANCHES OF ALIMENTARY INDUSTRY, 1928–1979

	1928	1950	1965	1979
Meat products (mill tonnes)	1.2	1.5	5.2	9.6
Butter (mill tonnes)	0.1	0.3	1.1	1.3
Milk products (mill tonnes)	n.d.	1.1	11.7	25.0
Fish products (mill tonnes)	0.8	1.8	5.8	9.4
Vegetable oils (mill tonnes)	0.4	0.8	2.8	2.8
Sugar (mill tonnes)	1.3	2.5	11.0	10.6
Tinned food (000 mill tins)	0.1	1.5	7.1	16.1

Alimentary industries as a group (Map 35A) are very widely dispersed and there are few urban centres of any size which do not have a representative of this sector of industry, which is strongly market oriented. Indeed, in recent years, food-processing plant have been set up in rural areas, not only in the smaller towns but also on the farms themselves. In view of the large number of producing centres, no attempt has been made to name them on the accompanying maps. At the same time, of course, the location of individual branches of food processing is influenced by the availability of the agricultural commodities on which they are based.

Map 35B Meat and dairy products

Meat processing and the production of dairy goods are among the most widespread of alimentary industries, occurring in nearly all agricultural areas. 25 years ago, the proportion of animal protein in the average Soviet diet was extremely low and the output of meat, butter and milk have all increased rapidly in recent years, though Soviet purchases of butter from the EEC, for example, suggest that demand still outstrips supply. Among the most important producing areas for all three commodities are the Ukraine (about 25 per cent) and the Centre, Volga and North Caucasus regions (some 8 or 9 per cent each).

Map 35C Fish processing

This, for very obvious reasons, has a highly distinctive location pattern, which includes not only coastal sites but also plant on major rivers and inland water bodies such as the Caspian and Aral seas and Lakes Balkhash and Baykal. Though these inland fisheries remain important, deep sea catches now predominate and Soviet vessels roam widely in the Atlantic, Pacific and Arctic Oceans. The USSR's total catch now exceeds 10 million tonnes annually, 14 per cent of the world total and second only by a small margin to that of Japan.

Map 35D Fruit and vegetable canning, wine and spirit making, tobacco processing

Not surprisingly, these activities are concentrated in southern areas of high value crop production – Moldavia, the Ukraine, North Caucasus, Transcaucasia and Central Asia. The output of canned food has increased nearly tenfold since 1950 and of current production about 30 per cent is in Central Asia, 30 per cent in Moldavia and the Ukraine and 25 per cent in the North Caucasus and Transcaucasia. The production of wine and spirits is a speciality of Moldavia, the Crimea, Georgia and Armenia.

Map 35E Flour milling and sugar refining

This map shows two activities which are associated predominantly with cereal cultivation and the growing of sugar beet in rotation with grain crops. However, since grain is more easily transported than flour, milling occurs mainly in major urban centres. Sugar production, on the other hand, is more clearly concentrated in the beet growing districts of the Ukraine, Kuban (North Caucasus region) and Central Asia. Of total sugar output, nearly 60 per cent comes from Moldavia and the Ukraine and about 10 per cent each from the Black Earth Centre and North Caucasus regions.

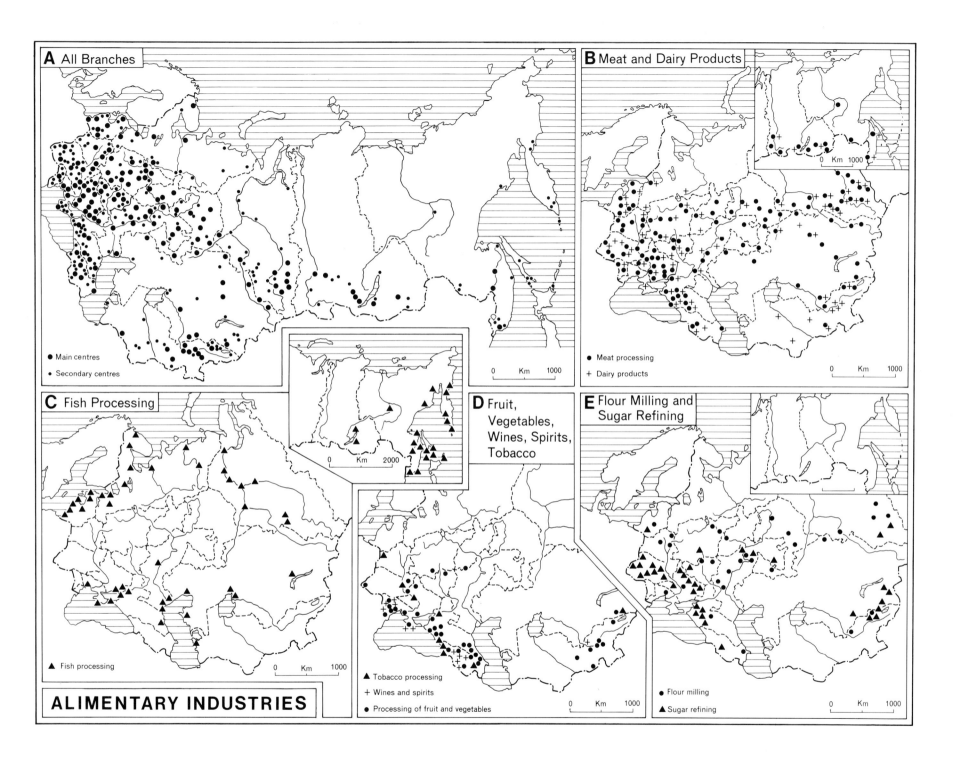

A All Branches

- ● Main centres
- ● Secondary centres

B Meat and Dairy Products

- ● Meat processing
- + Dairy products

0 Km 1000

C Fish Processing

- ▲ Fish processing

0 Km 2000

0 Km 1000

ALIMENTARY INDUSTRIES

D Fruit, Vegetables, Wines, Spirits, Tobacco

- ▲ Tobacco processing
- + Wines and spirits
- ● Processing of fruit and vegetables

0 Km 1000

E Flour Milling and Sugar Refining

- ● Flour milling
- ▲ Sugar refining

0 Km 1000

36 RAILWAYS

The vast size of the national territory, the great distances which separate raw materials, manufacturing centres and markets, and the development of a modern economy with a massive heavy industrial base are responsible for an enormous volume of freight movement within the USSR. Since the Revolution, such traffic has increased nearly fifty-fold, from 126 billion tonne-kilometres in 1913 to 6,200 billion tonne-kilometres in 1980. Over the same period, passenger movement has risen from 27 billion to 890 billion passenger-kilometres.

Throughout this period, railways have remained overwhelmingly predominant in the transport system of the USSR. (See Tables 12 and 13, page 78). In 1979 they were still responsible for 56 per cent of all freight movement and 39 per cent of the passenger traffic. These figures indicate a decline in the relative though not in the absolute position of the railways over the past twenty years, for in 1955 they were responsible for more than 80 per cent of both freight and passenger movement. This change reflects the increasing importance of road transport and, in the case of passenger movement, the airlines. Despite this decline in the railways' share of total traffic, the volume of freight movement by rail continues to rise rapidly; it increased by over 35 per cent between 1970 and 1980. The great bulk of movement by road is in any case over very short distances. While the average length of rail freight hauls continues to rise and is now in excess of 850 km, road freight consignments move on average less than 20 km. Freight traffic undoubtedly provides the great bulk of the railways' income, there being more than 10 tonne-kilometres of freight movement for every kilometre of passenger travel. The dominant role of the railways is a result of national transport policy and the Soviet Union has purposely avoided that competition between transport media which has been responsible for the decline of the railways in so many western countries.

The length of the rail network has doubled during the Soviet period, from about 70,000 km in 1913 to 142,000 km in 1980, and expansion continues. Much of the network in the European section of the country was constructed during the last fifty years of the Tsarist period, along with several important routes in the Asiatic regions, notably the Trans-Caspian and Trans-Aral lines to Central Asia and the Trans-Siberian to the Pacific. Railway building during the Soviet period, although it has included the construction of several important stretches in the European regions in order to shorten routes between major cities, has occurred mainly in a central zone stretching from the Urals to Lake Baykal, where economic development has been particularly intensive. Major elements here include the South Siberian and Middle Siberian routes, linking the Kuzbass and Karaganda with the Urals, the Turk-Sib railway linking the Central Asian area to the Trans-Siberian, several lines tapping the mineral riches of Kazakhstan and the improvement of regional networks in the Urals and West Siberia.

Despite the importance of the railways and the continuing expansion of the network, the latter is a very open one, with one kilometre of track for every 165 sq. km and 1600 people, which may be compared with 28 sq. km and 600 people in the USA. The layout of the network reflects the general distribution of population and economic activities, and there is a marked contrast between the situation in the European part of the country, with a fairly complete network, and that in the Asiatic regions, where the system comprises a series of major trunk routes with numerous branch lines to individual centres of economic activity.

Extension of the network in recent years has been mainly in the form of new branch lines into new areas of mineral extraction, mainly in Siberia, but one major project is now under way, namely the Baykal-Amur Mainline (BAM). Although originally planned in the 1930s, construction actually began in 1974 and completion is scheduled for 1983. This 3000 km line, which parallels the Trans-Siberian some 200–300 km further north, is designed to facilitate the exploitation of the natural resources of East Siberia and the Far East and to strengthen the link between Europe and the Pacific. The strategic vulnerability of the Trans-Siberian, which for much of its length runs close to the border with China, is an additional stimulus. Another major piece of construction, announced in 1976, is the building of a railway across the Caucasus which, though short, will be a considerable feat of engineering.

The rather limited route length and the great volume of freight movement result in the heaviest loadings in the world, with an average annual movement of 23 million tonne-kilometres for every kilometre of track. A few major bulk commodities account for more than three-quarters of the traffic; on a tonne-kilometre basis the main items are coal and coke (17%), petroleum products (15%), ores and metals (15%), mineral building materials (13%), timber (10%), grain (4%) and mineral fertilisers (3%). Since much of this traffic is concentrated on a few trunk routes, some very heavy loadings occur. The busiest lines of all, carrying well over 50 million tonnes annually, are those associated with the major industrial combines of the Urals-Kuzbass and Donbass-Dnepr Bend. Densities of 25–50 million tonnes occur on lines linking Moscow with the Donbass and Urals, on the Karaganda-Magnitogorsk section and on the Novosibirsk-Irkutsk section of the Trans-Siberian. Lines with loadings of 15–20 million tonnes join Leningrad, the western Ukraine and the Caucasus with the areas already mentioned.

The heavy and increasing pressure on these major routes has led to large-scale investment in improvements to the system such as better signalling, the replacement of steam by diesel and electric traction and double tracking. About a quarter of the route length, including all the heavily-utilised lines already mentioned, is now electrified. Continuing heavy investment in the railway system is an integral part of Soviet economic plans.

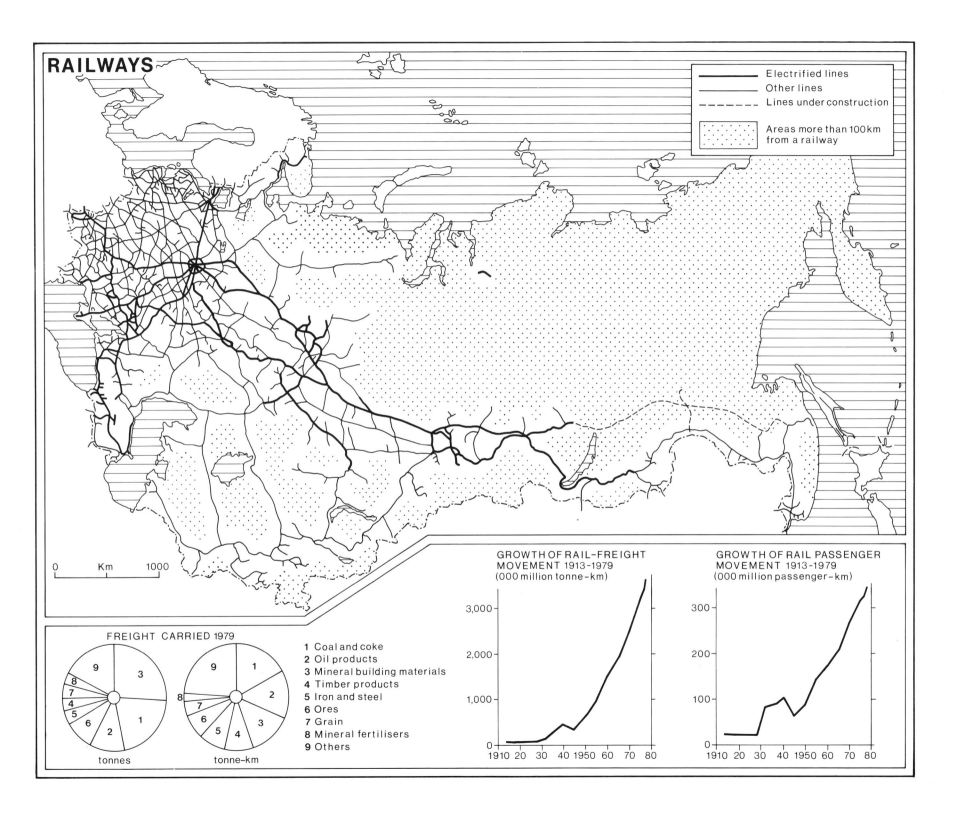

RAILWAYS

Legend:
- Electrified lines
- Other lines
- Lines under construction
- Areas more than 100km from a railway

0 Km 1000

FREIGHT CARRIED 1979

tonnes

tonne-km

1 Coal and coke
2 Oil products
3 Mineral building materials
4 Timber products
5 Iron and steel
6 Ores
7 Grain
8 Mineral fertilisers
9 Others

GROWTH OF RAIL-FREIGHT MOVEMENT 1913-1979
(000 million tonne-km)

3,000

2,000

1,000

0

1910 20 30 40 1950 60 70 80

GROWTH OF RAIL PASSENGER MOVEMENT 1913-1979
(000 million passenger-km)

300

200

100

0

1910 20 30 40 1950 60 70 80

37 ROADS

As the accompanying tables show, the roads play a relatively minor role in the internal transport system of the Soviet Union, despite a considerable increase in the use of road transport in recent years. This is particularly true in the case of freight movement. Though road freight turnover has tripled since 1965, it is still less than seven per cent of the Soviet total and only about one-eighth of the movement by rail. In volumetric terms, roads carried 21.5 billion tonnes of freight in 1978, compared with 3.6 billion tonnes carried by rail, but the average distance of a road shipment was only 16.5 km as against more than 850 km by rail. The vast bulk of road freight movements is over very short distances to or from the nearest railhead and long hauls occur only where railways are lacking, as for example to the north of the Trans-Siberian railway; in such areas the actual volume of movement is small. In recent years, an increasing proportion of Soviet foreign trade and of internal traffic within the European regions has been by road, but this still represents only a minute fraction of the total.

Over the past 20 years, passenger movement by road has increased very rapidly and, at 43 per cent of the total, is slightly greater than that by rail. The 376 billion passenger-kilometres recorded, however, involves 41.3 billion individual journeys (157 per head of the population) with an average length of less than 10 km; the great bulk of recorded movement is by public transport systems within urban areas or from rural areas to nearby towns. There is a small but growing movement by bus between the cities of the European regions, but most long-distance journeys are by rail or air.

Map 37 shows the general layout of the road network, which is closest in the more heavily populated sections of the European USSR and skeletal or non-existent in the Asiatic regions. Hard-surfaced roads now total 690,000 km, four times the 1960 figure but still the equivalent of only one km per 32 sq km of territory and there are only 315,000 km of improved road with a cement or asphalt surface. As the inset map shows, there are pronounced regional contrasts in the density of the network with values above one km of road per 10 sq km in the Baltic republics, the western Ukraine, the Transcaucasus and the Moscow oblast but less than one km per 100 sq km of territory over practically the whole of Siberia and much of Central Asia.

The Soviet Union stands alone among developed countries in the limited role accorded to road transport. Though now in a phase of fairly rapid growth, road transport is unlikely, in the foreseeable future, to replace rail transport in the way in which it has done so in western or North America.

TABLE 12 FREIGHT TURNOVER (thousand million tonne-kilometres) (Figures in brackets are percentages of the USSR total)

	1928	1950	1965	1979
Rail	93.4 (78.2)	602.3 (84.4)	1950.2 (70.6)	3349.3 (56.0)
Road	0.2 (0.2)	20.1 (2.8)	143.1 (5.2)	407.9 (6.8)
Sea	9.3 (7.8)	39.7 (5.6)	388.8 (14.1)	851.1 (14.2)
Waterway	15.9 (13.3)	46.2 (6.5)	133.9 (4.8)	232.7 (3.9)
Pipeline	0.7 (0.6)	4.9 (0.7)	146.7 (5.3)	1140.7 (19.1)
Air	0.0 (0.0)	0.1 (—)	1.3 (—)	2.9 (—)
Total	119.5 (100.0)	713.3 (100.0)	2764.0 (100.0)	5984.6 (100.0)

TABLE 13 PASSENGER MOVEMENT (thousand million passenger-kilometres) (Figures in brackets are percentages of the USSR total)

	1928	1950	1965	1979
Rail	24.5 (90.4)	88.0 (89.5)	201.6 (55.0)	335.3 (38.5)
Road	0.2 (0.7)	5.2 (5.3)	120.5 (32.9)	376.0 (43.2)
Sea	0.3 (1.1)	1.2 (1.2)	1.5 (0.4)	2.5 (0.3)
Waterway	2.1 (7.7)	2.7 (2.7)	4.9 (1.3)	5.8 (0.7)
Air	0.0 (0.0)	1.2 (1.2)	38.1 (10.4)	151.0 (17.3)
Total	27.1 (100.0)	98.3 (100.0)	366.6 (100.0)	870.6 (100.0)

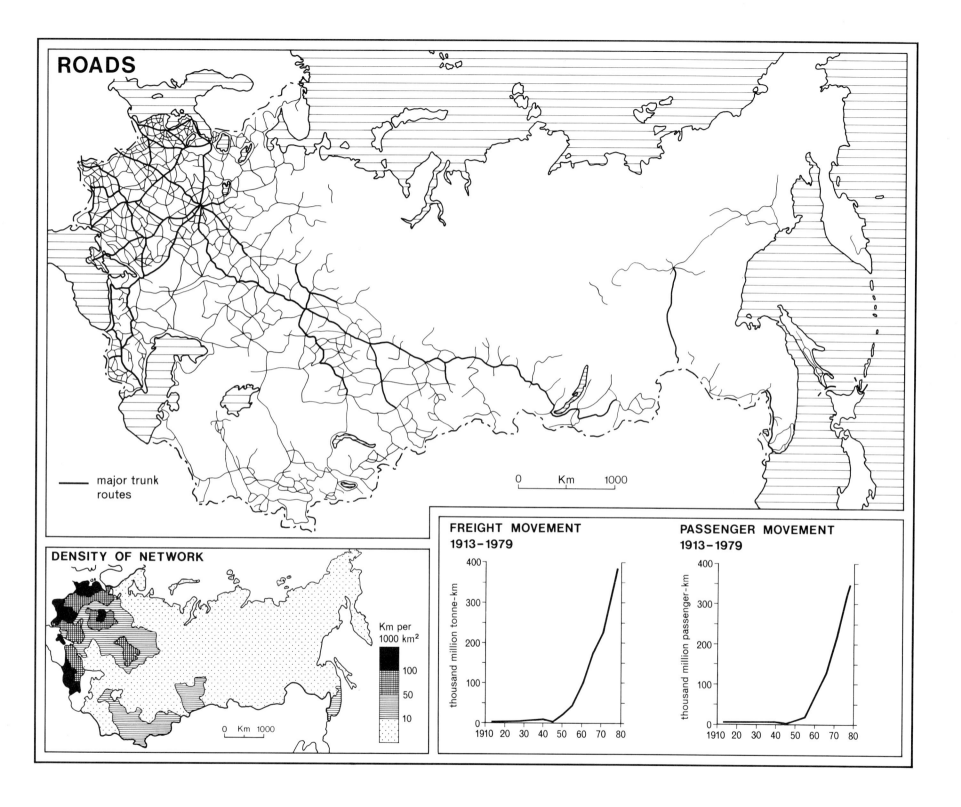

ROADS

— major trunk routes

0 Km 1000

DENSITY OF NETWORK

Km per 1000 km²

100
50
10

0 Km 1000

FREIGHT MOVEMENT 1913–1979

thousand million tonne-km

400
300
200
100
0

1910 20 30 40 50 60 70 80

PASSENGER MOVEMENT 1913–1979

thousand million passenger-km

400
300
200
100
0

1910 20 30 40 50 60 70 80

38 WATERWAYS

While Map 4 shows all natural waterways of the USSR, Map 38 shows only those waterways, both natural and artificial, identified in Soviet texts and atlases as navigable.

Inland waterways played a major role in the internal transport system of the Russian state from its earliest days. Indeed, until the advent of the railway in the middle of the nineteenth century, they were the most important form of transport over medium and long distances. The European rivers, their headwaters separated by short, low porterages, carried traffic between the Baltic/Scandinavian world and the Black Sea/Mediterranean from at least the tenth century A.D. and the Slav principalities of the early medieval period derived considerable wealth from this traffic. In addition, links via the Caspian to Turkestan and even as far east as China were important to the Russians from the sixteenth century or earlier, while Russian exploration of Siberia depended to a considerable extent upon movement along the waterways, particularly along the east-west tributaries of the great Siberian rivers.

In modern times, inland waterways, especially those of European Russia, have continued to play an important role. On the eve of the First World War they were responsible for about one third of all internal freight movement within the Russian Empire; the total of some 28,900 tonne-kilometres in 1913 represented the shipment of 35,100,000 tonnes of freight over an average distance of 823 km. Of the materials carried, the most important were timber (31 per cent), grain (17 per cent), petroleum products (15 per cent) and mineral building materials (4 per cent).

In the period since the Revolution there has been large-scale investment in improvements to the inland waterway system, involving the construction of new canals and the rebuilding of old ones together with the dredging and widening many rivers. Major hydroelectric schemes (see Map 25) have often involved incidental improvements to navigation along the rivers concerned.

The amount of freight movement has increased nearly eight-fold, reaching 230,000 million tonne-kilometres in the late 1970s while the actual volume carried has risen thirteen-fold to about 500 million tonnes. The average distance of a waterway shipment, however, has declined to about 460 km.

With developments in rail, pipeline and road transport systems, the waterways have been increasingly concerned with medium-distance bulk movement of low-value commodities, especially mineral building materials, which now constitute 60 per cent of all freight carried, and timber (18 per cent), though considerable quantities of petroleum products, coal, coke, grain, ores and metals are still moved by river and canal. Despite the large volumetric increase, however, the waterways' share of total freight movement has shrunk to about 4 per cent.

Inland waterways play only a minor role in passenger transport, despite the introduction of high-speed vessels on several major routes. Their share of total passenger movement has fallen from about 5 per cent 60 years ago to 0.8 per cent in the late 1970s. Nevertheless this seemingly insignificant fraction amounts to about 6,000 million passenger-kilometres and involves 145,300,000 million individual journeys over an average distance of only 42 km.

The most highly developed system is that based on the Volga, which is responsible for 50–60 per cent of freight movement. Flowing as it does across several environmental zones, the Volga has long been a medium of exchange for commodities produced along its course, but the river serves a wider function. In its upper reaches, it flows within 150 km of Moscow and its tributary, the Oka is joined by the Moscow river which flows through the heart of the city. These natural links have been strengthened by the construction of the Moscow Canal (3)*. Further east, the Kama tributary taps traffic from the west flank of the Urals, while down-stream from the Kama-Volga confluence the river has been converted to a series of man-made lakes, ponded back behind a series of hydro-electric barrages, which have increased its carrying capacity while leading to delays at the various locks. The ramifications of the Volga

system, however, extend far beyond its own catchment area. To the north, canals linking the Volga to the Sukhona-Dvina system and to Lake Onega, whence there are links to the White Sea and Baltic, were built in the early nineteenth century and have been improved as part of the modern Volga-Baltic Waterway (2) completed in 1964. The White Sea-Baltic Canal (1) was built in the interwar period. In the south, the Volga-Don Canal (5), completed in 1952, provides connections with the eastern Ukraine and a link between the Volga and the Black Sea, while the Caspian provides links with Transcaucasia and Central Asia.

Further west, the Dnepr provides another north-south route allowing the entry of raw materials to the Ukrainian industrial zone, but the amount of traffic is much less than on the Volga. The Dnepr-Bug Canal (4), providing a link with Poland, is little used (its main function is to help drain the Pripet Marshes), and no link has been planned to the Western Dvina, which is navigable 500 km inland from the Baltic.

In Central Asia, two rivers and the Kara Kum irrigation canal are recorded as navigable but carry only a small volume of local traffic.

The major rivers of Siberia – Irtysh, Ob, Yenisey, Lena, Aldan, Kolyma – have large carrying capacities but are little used. Nature has decreed they flow northward, while the prime need of the region is movement between east and west. During the short season for which they are open, they carry timber and minerals both northwards to the Arctic Sea route and southwards to transhipment points on the Trans-Siberian railway. The Amur system, though it flows through the more developed parts of the Far East region, is also little used, partly because of the relatively low level of economic activity in the Far East, partly because it is paralleled by the Trans-Siberian railway.

All inland waterways have disadvantages which weigh against successful competition with other transport media, notably the long winter freeze, the indirect routes which they provide and the slow speed of freight vessels.

* A number in brackets refers to the key number on the map.

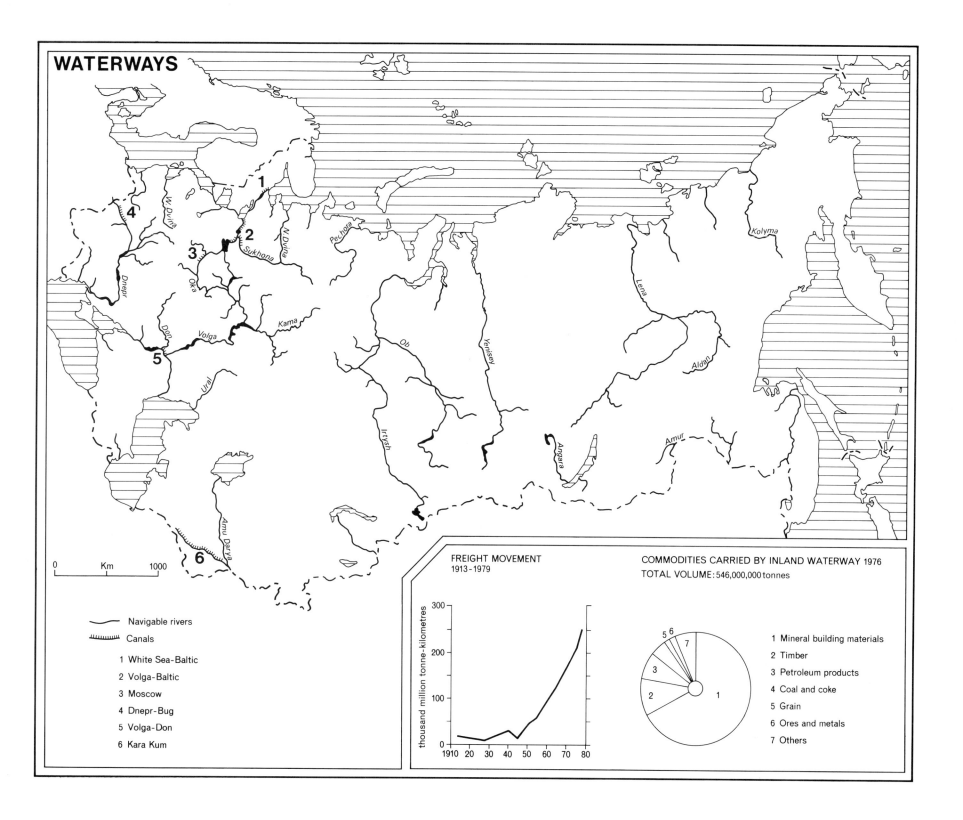

WATERWAYS

Dnepr
W.Dvina
N.Dvina
Sukhona
Pechora
Kolyma
Lena
Oka
Kama
Don
Volga
Aldan
Ob
Yenisey
Ural
Irtysh
Angara
Amur
Amu Darya

1
2
3
4
5
6

0   Km   1000

Navigable rivers
Canals

1 White Sea-Baltic
2 Volga-Baltic
3 Moscow
4 Dnepr-Bug
5 Volga-Don
6 Kara Kum

FREIGHT MOVEMENT
1913-1979

thousand million tonne-kilometres

300

200

100

0

1910 20 30 40 50 60 70 80

COMMODITIES CARRIED BY INLAND WATERWAY 1976
TOTAL VOLUME: 546,000,000 tonnes

5 6
7
3
2
1

1 Mineral building materials
2 Timber
3 Petroleum products
4 Coal and coke
5 Grain
6 Ores and metals
7 Others

39 FOREIGN TRADE

A striking feature of the past 25 years has been a massive increase in the volume of Soviet foreign trade, some aspects of which are illustrated in the accompanying diagrams. Total trade turnover (A), which in 1950 was a mere 2,925 million roubles, reached 22,100 million roubles in 1970 and 80,290 million roubles in 1978. Owing to successive revaluations of the rouble (from $1.1111 in 1950 to $1.4628 in 1979), growth has been even more impressive in dollar terms: trade turnover rose from $3,250 million in 1950 to $117,448 million in 1978, a 36-fold increase. Over the same period, per capita turnover increased from 16.3 roubles ($18.1) to 306 roubles ($448).

Over the period as a whole, the Soviet Union has maintained a favourable trade balance equivalent to about two per cent of turnover; in individual years the size of this balance has fluctuated considerably and, during the 1970s, when turnover has increased particularly rapidly, the amplitude of these fluctuations has also increased. The 1970s have included years of major trade deficit (e.g. 1975: deficit 2,637 mill. roubles; $3,655 mill.) as well as years with a very large surplus (e.g. 1979: 4,562 mill. roubles; $6,673 million). For the period 1970–79 as a whole, however, a favourable balance has continued with exports totalling 236,599 million roubles ($320,715 mill.) and imports valued at 227,392 million roubles ($308,096 mill.), an average annual surplus of 921 million roubles ($1,262 mill.).

As Figure 39B shows, the period under review has also witnessed significant changes in the direction of Soviet foreign trade. Whereas the proportion of total trade turnover involving the centrally planned economies was 81 per cent in 1958, twenty years later it had fallen to 60 per cent, with a compensating rise in trade with the capitalist world from 19 to 40 per cent. The share of the less developed market economies was much the same at both dates; practically all the increase in trade with the capitalist world was accounted for by the growth in trade with the developed market economies of western Europe, Japan and North America. These trends have been particularly strong on the import side; Soviet imports from the developed market economies rose, between 1958 and 1978, from 12.5 to 35 per cent of the total. In value terms this represents an increase from 489 million roubles ($544 mill.) to 12,093 million roubles ($17,690 mill.), a 25–30 fold increase. The favourable balance maintained by Soviet trade in general has not eliminated the problem of hard currency shortages which still limit the Soviet Union's ability to import from the west.

Figure 39C gives details of trade with the Soviet Union's main partners, both communist and non-communist and their share of 1979 imports and exports is shown in 39D. Whereas in the 1960s trade with every non-communist country was much less than that with communist states, during the 1970s the gap has narrowed. Nevertheless, the USSRS COMECON partners in Europe, along with Cuba, continue to dominate the pattern. In 1977 the same five leading partners – the German Democratic Republic, Poland, Czechoslovakia, Bulgaria and Hungary – took 41 per cent of Soviet exports and provided 43 per cent of Soviet imports. Among the capitalist countries, the leading exporters to the USSR were the USA (6.6 per cent), Federal Germany (5.9 per cent) and Japan (4.4 per cent), while the leading capitalist importers of Soviet goods were Federal Germany (4.4 per cent), Finland (3.5 per cent), France (3.4 per cent) and Italy (3.0 per cent). Particularly interesting is the case of the United States. From a situation in the early 1960s, when trade between the superpowers was negligible, there has been a dramatic change to one in which the USA is a major exporter to the USSR but receives few Soviet imports. This is due mainly to the fact that the USA has made up much of the Soviet Union's grain deficit in years when harvests have been poor, a fact which is also responsible for the wide annual fluctuations in both the volume of trade between the two countries and the USSR's overall trade balance.

Figure 39E indicates the commodity structure of Soviet imports and exports in 1965 and 1979, the latter being fairly characteristic of the situation in recent years. On the import side, foodstuffs were a major item, accounting for 22 per cent of all imports, but by far the most important category was machinery and transport equipment, 38 per cent of all imports. 42 per cent of all exports were of fuels and energy, predominantly oil and natural gas. As the diagram shows, this had been a relatively minor element in the mid-1960s and the build-up of Soviet oil and gas exports has been a major feature of the 1970s. From 1970 to 1978 inclusive, the USSR exported some 800 million tonnes of oil (about 20 per cent of her total output) of which nearly a third went to hard currency areas. Natural gas exports built up rapidly in the 1970s, reaching 20 billion cu m (5 per cent of total output) in 1979, and more than 40 per cent of this went to western Europe. Oil and gas exports are a major item in assisting the USSR to purchase high technology goods from the capitalist world and are essential to her economic development. The sizeable export of manufactured goods, however, went mainly to the Communist states.

Although the USSR has a high degree of self-sufficiency and foreign trade is a minor element in her economy when compared with its role in the economies of, say, the UK, Japan or West Germany, the situation has changed considerably in the post-war period. The establishment of socialist states in eastern Europe and their collaboration within COMECON has led to increased mutual dependence between the USSR and her neighbours. More recently, trade with the capitalist world has played an increasing role in Soviet industrial development and in this sense east and west have to some degree become more dependent on each other.

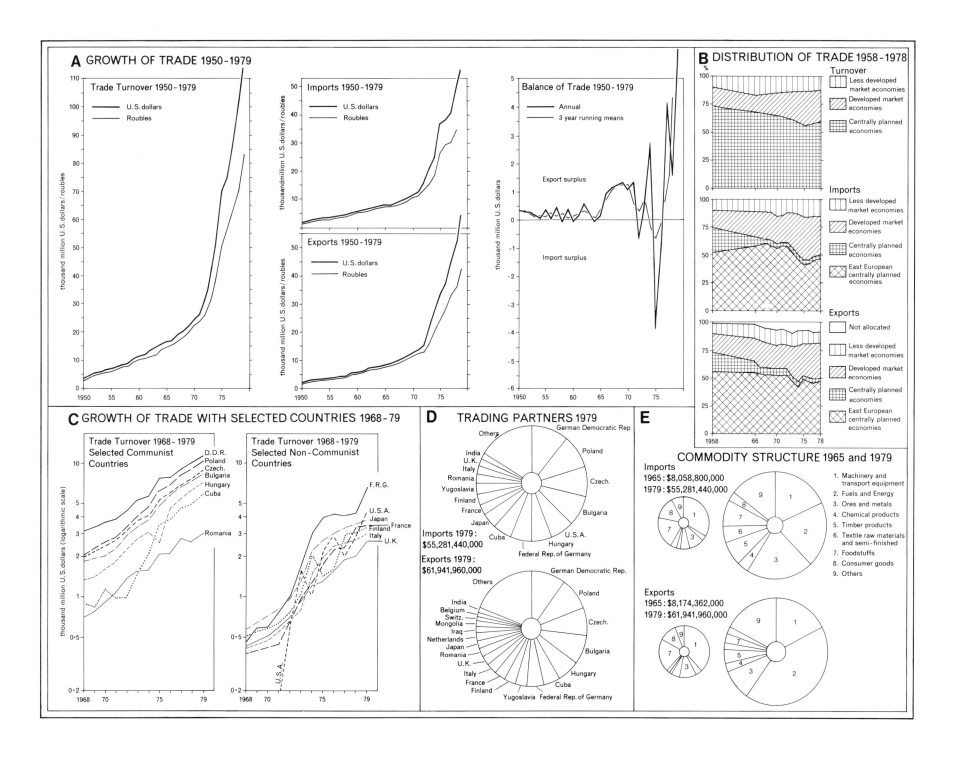

A GROWTH OF TRADE 1950-1979

Trade Turnover 1950-1979
— U.S. dollars
— Roubles

Imports 1950-1979
— U.S. dollars
— Roubles

Exports 1950-1979
— U.S. dollars
— Roubles

Balance of Trade 1950-1979
— Annual
— 3 year running means

Export surplus

Import surplus

B DISTRIBUTION OF TRADE 1958-1978

Turnover
- Less developed market economies
- Developed market economies
- Centrally planned economies

Imports
- Less developed market economies
- Developed market economies
- Centrally planned economies
- East European centrally planned economies

Exports
- Not allocated
- Less developed market economies
- Developed market economies
- Centrally planned economies
- East European centrally planned economies

C GROWTH OF TRADE WITH SELECTED COUNTRIES 1968-79

Trade Turnover 1968-1979 Selected Communist Countries
D.D.R.
Poland
Czech.
Bulgaria
Hungary
Cuba
Romania

Trade Turnover 1968-1979 Selected Non-Communist Countries
F.R.G.
U.S.A.
Japan
France
Finland
Italy
U.K.
U.S.A.

D TRADING PARTNERS 1979

Others
India
U.K.
Italy
Romania
Yugoslavia
Finland
France
Japan
Cuba
Hungary
Federal Rep. of Germany
U.S.A.
Bulgaria
Czech.
Poland
German Democratic Rep

Imports 1979:
$55,281,440,000

Exports 1979:
$61,941,960,000

Others
India
Belgium
Switz.
Mongolia
Iraq
Netherlands
Japan
Romania
U.K.
Italy
France
Finland
Yugoslavia
Federal Rep. of Germany
Cuba
Hungary
Bulgaria
Czech.
Poland
German Democratic Rep.

E COMMODITY STRUCTURE 1965 and 1979

Imports
1965 : $8,058,800,000
1979 : $55,281,440,000

Exports
1965 : $8,174,362,000
1979 : $61,941,960,000

1. Machinery and transport equipment
2. Fuels and Energy
3. Ores and metals
4. Chemical products
5. Timber products
6. Textile raw materials and semi-finished
7. Foodstuffs
8. Consumer goods
9. Others

PART 4 REGIONS

40 KEY TO THE REGIONAL MAPS

The nine maps (41—49) on pages 89—105 show selected areas of Soviet territory and together cover the whole of the USSR. The division of the Soviet Union into a set of areas for regional description presents a variety of problems, and these are intensified when the regional accounts are to be based on a set of maps. That a set of 'regional' maps should form the concluding section of this volume seems obvious enough — it is only by the study of such maps and the accompanying texts that the reader can gain some idea of the way in which the varied elements of the geography of the USSR, both physical and human, portrayed in the rest of this volume come together and interact to impart a distinctive character to each region.

Ideally, perhaps, the maps in this section would all have been drawn not only with a common set of symbols but also to a common scale. Such a procedure is clearly impracticable, not least because of the basic geography of the USSR itself, particularly its vast size and the extremely uneven distribution of its population and economic activities. To show the complex economic geography of the Ukraine (Map 43), for example, requires a map on a scale not less than 100 kilometres to the centimetre; maps on a similar scale covering the whole of Siberia (Map 46) would have covered some 15 pages, many of which would have been virtually blank; conversely, using the same scale as that finally adopted for the Siberian regions, the Donets-Dnepr industrial complex in the Ukraine (Map 43) would have been compressed to a size of 2.3 × 1.8 cm. Consequently, a variety of scales have been used on the general principle that, the more densely populated and intensively developed the area involved, the larger the scale required. Thus the regional maps range in scale from 62.5 km to the cm for the Donbass (Map 43D) and the Moscow district (Map 42B) to 350 km to the cm for the empty lands of northern Siberia (Map 46A).

Further problems are involved in the choice of a suitable set of regions to be mapped individually. Given the limitations of the space available, this had to be such as to show the entire country in a minimum number of pages yet at the same time to give as much detail as possible in each case. Nine maps, each covering one-ninth of the national territory, would have been the worst possible solution; in the event, the area covered by a single page ranged from 541,200 sq. km or one fortieth of the USSR (Map 47) to 12,765,900 sq. km (Map 46), well over half the land area.

It will also be observed that the official set of Major Economic Regions (see p. 24) has been used as a basis. These, too, are far from ideal; some of them have great internal diversity and little coherence, economic or otherwise, though in most cases there is some degree of unity, imparted by the physical environment, the ethnic characteristics of the population and/or the nature of their economic development. They are certainly meaningful to many Soviet geographers, who have used them as a regional framework for their own text books and atlases, and they are at least no less satisfactory than any one of many possible sets of a similar number of *ad hoc* regions. The Major Economic Regions, however, are 19 in number and range in size from 167,700 sq. km (Black Earth Centre)* to 6,215,900 sq. km (the Far East). To map each separately would have produced an excessive number of maps and would, in any case, run up against the scale problem already discussed. Of the nine regional maps finally decided upon, four are each of a single region (Maps 44 Volga, 45 Urals, 48 Kazakhstan, 49 Central Asia) and the other five cover what it is hoped are fairly logical combinations of regions (the North-west, Baltic and Belorussia, Map 41; the Centre, Volga-Vyatka and Black Earth Centre, 42; the three Ukrainian regions and Moldavia, 43; the three Siberian regions, 46; and the North Caucasus and Transcaucasia, 47).

To map each of these nine areas in a precisely similar manner would have been unsatisfactory and rather dull; what are required are maps emphasising the salient points of each. Again, to show all the important features of each area on a single map would have produced maps so complex as to be almost incomprehensible; thus each page contains several maps of the chosen area. Some of these are common to all sets; in every case there is an initial map (A) showing rivers, administrative divisions, railways and all towns with more than 50,000 inhabitants, which acts as a reference framework for the rest of the set. Subsequent maps show, in every case, though in slightly different combinations, the more basic elements of the industrial economy — energy and mineral production and metallurgical activities. In addition, there are sometimes maps unique to a set, which either enlarge an area of particular complexity, like the Moscow district (Map 42B) or the Donbass (Map 43D), or portray some feature of special significance such as the environmental zones of the Volga region (Map 42), relief in Caucasia (Map 47), oil and gas production in the Tyumen oblast (Map 46B).

The symbols used are largely self-explanatory and require little comment. It should be emphasised, however, that each symbol depicts the location of a particular activity and not every individual mine or factory. Under metallurgy, for example, a location marked by one of the symbols for steel plant may well have more than one such plant — this is why there are three symbols for ferro-alloy production, which may occur on its own or in a town which also has an integrated steel plant or a steel mill. Similarly, the symbols for the various types of mineral extraction indicate localities; each may represent several mines. In the case of electricity generation, one symbol may represent a number of separate thermal generating stations. Hydro-electric plant, on the other hand, are shown individually, save in the case of 'cascades' of plant along a short stretch of river, where a slightly different symbol is used with a number alongside to indicate the number of stations.

Finally, the inset on the 'Key to Regions' map indicates the set of Major Economic Regions and their combinations in Maps 41—49; the bar graphs show the area and population of each region.

* One of the 19 units — Moldavia — covers only 33,700 sq. km, but is not, strictly speaking, a Major Economic Region.

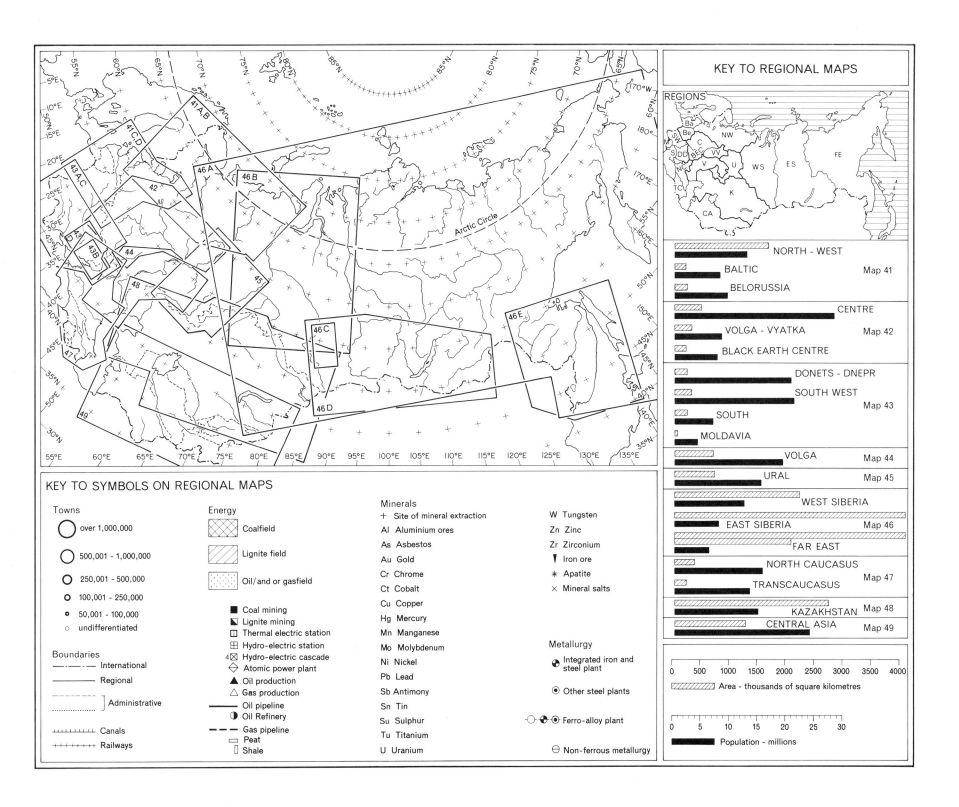

KEY TO REGIONAL MAPS

REGIONS

NORTH - WEST
BALTIC — Map 41
BELORUSSIA

CENTRE
VOLGA - VYATKA — Map 42
BLACK EARTH CENTRE

DONETS - DNEPR
SOUTH WEST — Map 43
SOUTH
MOLDAVIA

VOLGA — Map 44

URAL — Map 45

WEST SIBERIA
EAST SIBERIA — Map 46
FAR EAST

NORTH CAUCASUS — Map 47
TRANSCAUCASUS

KAZAKHSTAN — Map 48
CENTRAL ASIA — Map 49

0   500   1000   1500   2000   2500   3000   3500   4000
▨ Area - thousands of square kilometres

0   5   10   15   20   25   30
■ Population - millions

KEY TO SYMBOLS ON REGIONAL MAPS

Towns
◯ over 1,000,000
◯ 500,001 - 1,000,000
◯ 250,001 - 500,000
○ 100,001 - 250,000
⊙ 50,001 - 100,000
○ undifferentiated

Boundaries
—·—·— International
———— Regional
– – – –] Administrative
············
+++++++ Canals
++++++++ Railways

Energy
▨ Coalfield
▨ Lignite field
▨ Oil/and or gasfield

■ Coal mining
◣ Lignite mining
⊟ Thermal electric station
⊞ Hydro-electric station
4⊠ Hydro-electric cascade
◇ Atomic power plant
▲ Oil production
△ Gas production
—— Oil pipeline
◑ Oil Refinery
– – – Gas pipeline
▭ Peat
▯ Shale

Minerals
+ Site of mineral extraction
Al Aluminium ores
As Asbestos
Au Gold
Cr Chrome
Ct Cobalt
Cu Copper
Hg Mercury
Mn Manganese
Mo Molybdenum
Ni Nickel
Pb Lead
Sb Antimony
Sn Tin
Su Sulphur
Tu Titanium
U Uranium

W Tungsten
Zn Zinc
Zr Zirconium
▮ Iron ore
✳ Apatite
✕ Mineral salts

Metallurgy
◕ Integrated iron and steel plant
◉ Other steel plants
○ ◕ ◉ Ferro-alloy plant
⊖ Non-ferrous metallurgy

87

41 THE NORTH-WEST, BALTIC AND BELO- RUSSIAN REGIONS

These three regions are displayed together mainly for convenience, though there is a certain degree of unity within each of the two sections mapped separately, the Leningrad, Pskov and Novgorod oblasts having more in common with the Baltic and Belorussian republics than with the rest of the North-west. Together, the North-west, Baltic and Belorussian regions have an area of 2,059,500 sq km (9.2 per cent of the Soviet total) and a population of 31 million (11.8 per cent) but more than 80 per cent of their inhabitants live in the southern districts shown on a larger scale (Maps C and D), which constitute about a third of the area mapped.

The more northerly section (Maps A and B) functions primarily as a source of energy and raw materials. Mineral fuels are produced mainly in the Komi ASSR, which contains both the Pechora coalfield and the Komi-Ukhta oil and gas deposits. Coal mining began at Vorkuta in the 1930s and was much expanded during World War II when the field was linked to the European rail network by the line to Kotlas. Mining in this Arctic district is difficult, reserves are limited and annual output is unlikely to expand beyond the present modest 25–30 million tonnes. Coal is supplied mainly to the industrial areas of the North-west, especially Leningrad, coking coal going mainly to the integrated iron and steel plant opened at Cherepovets, on the Rybinsk reservoir, in 1955. The Komi-Ukhta oilfield also began production in the 1930s, but output remained low until the mid-1950s. The 1979 output of 19 million tonnes of oil and 18,000 million cu m of gas were only 3.2 and 4.4 per cent of the respective Soviet totals. The gas pipeline shown on Map B now carries gas from West Siberia.

In the north-west, the Murmansk oblast is the scene of a variety of important mining operations. One of the first items to be exploited in this area was apatite, a mineral used in the production of phosphatic fertilisers (see p. 64). Mining began at Kirovsk in the 1930s and was much expanded after World War II with the opening of a second site at Apatity. The war was also followed by the annexation of the Petsamo district, with its nickel mines, from Finland and a further source of nickel was developed later near Monchegorsk. Iron ore mining began at Olenogorsk in 1955 and Kovdor in 1962; concentrate from these sources is sent to Cherepovets.

This western zone is also of considerable importance in the aluminium industry. The first developments occurred in the early 1930s when bauxite mined at Boksitogorsk (Leningrad oblast) was transformed to alumina nearby and the alumina reduced to aluminium at the Volkhov hydro-electric site. This bauxite source is now nearly exhausted but replacements have been found in the form of nephelite, a by-product of apatite production at Kirovsk and, more recently, a new bauxite source at Plesetsk in the Onega valley. Much of the energy consumed in these mining and metallurgical activities is derived from hydro-electric plant in Karelia and the Murmansk oblast. Numerous small rivers – Voronya, Tuloma, Kovda-Kuma, Kem, Vyg – descend from the Finno-Karelian lake plateau and 'cascades' of small generating plant have been built along them. In all there are more then a score of such stations, but their total generating capacity of less than 2 million kW is below that of most individual plant along major rivers like the Dnepr, Volga and those in Siberia.

The southern section (Maps C and D), in contrast, is poor in industrial resources and, with the exception of the Leningrad district, was an industrial backwater until after World War II – indeed the Baltic republics and much of Belorussia were not finally incorporated in the USSR until 1945. This section is devoid of coal and the one small oilfield at Rechitsa in Belorussia is a very minor producer. In the absence of local coal and oil, there has been intensive exploitation of peat, which supports major thermal generating plant at Khodino, Bereza and Svetlogorsk in Belorussia and at Riga and Vilnyus in the Baltic republics, as well as several smaller stations. Northern Estonia is one of the few areas in the USSR where oil shale is extracted on a large scale. In addition, hydro-electric stations have been built on the Narva, Neman and Western Dvina rivers.

Despite these developments, the indigenous energy base remains weak but, over the past 25 years, the energy position has been transformed by the arrival of gas and oil through the pipeline network and this has permitted accelerated industrialisation. Heavy industry is poorly represented, being confined to steel mills at Liyepaya on the Baltic coast and in the Leningrad district – in the latter case there are heavy engineering plant (see p. 60) – but light industries of various kinds have been established in many of the regions' cities and have helped to absorb surplus labour from the poorer rural areas. Among the most important centres are Minsk (machine tools, motor vehicles, tractors, electrical equipment), Riga (railway rolling stock, electrical equipment, buses, ships), Vilnyus (machine tools and agricultural machinery), Tallin (electrical equipment, ships) and Kaliningrad (railway rolling stock, shipbuilding). Textiles are also important, based originally on flax but more recently including synthetic textiles, part of the development of the chemical industries assisted by the products of the Novopolotsk and Mozyr refineries.

Thus these regions, which 25 years ago were predominantly agricultural, have experienced rapid urban-industrial growth and now make a considerable contribution to the manufacturing as well as to the extractive sector of Soviet industry.

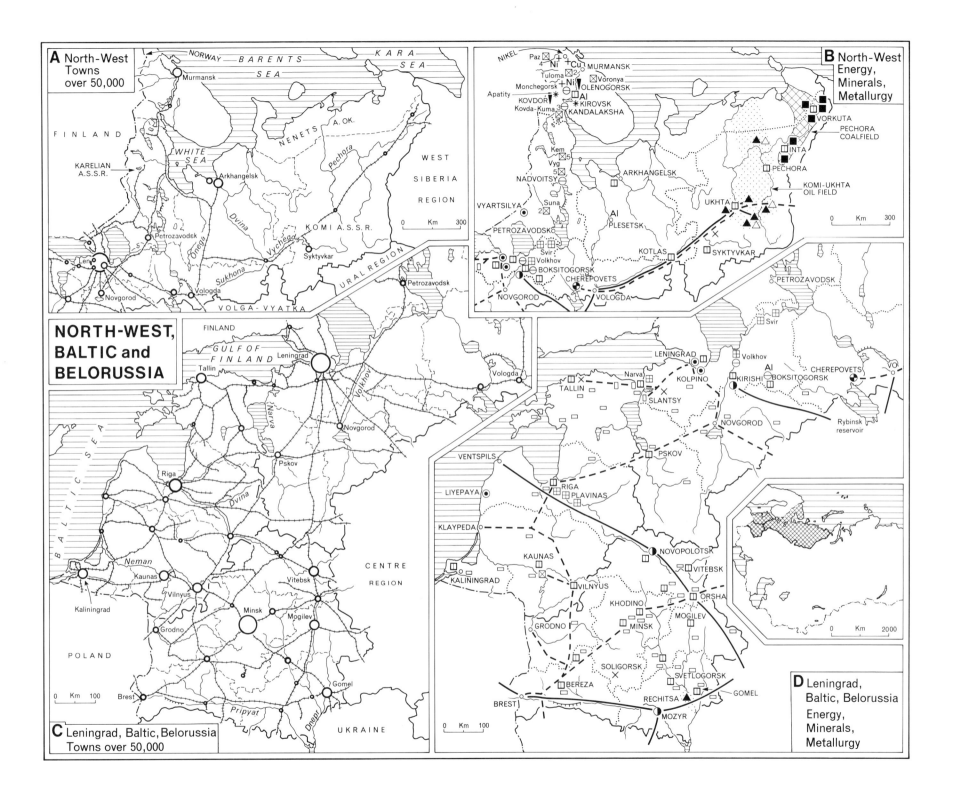

A North-West Towns over 50,000

NORWAY
BARENTS SEA
KARA SEA
Murmansk
FINLAND
WHITE SEA
NENETS A. OK.
KARELIAN A.S.S.R.
Arkhangelsk
Pechora
WEST SIBERIA REGION
Onega
Dvina
Vychegda
KOMI A.S.S.R.
Len
Petrozavodsk
Sukhona
Syktyvkar
URAL REGION
Vologda
Novgorod
VOLGA-VYATKA
0 Km 300

B North-West Energy, Minerals, Metallurgy

NIKEL
Paz
Ni +Cu MURMANSK
Tuloma
Monchegorsk Voronya
Ni OLENOGORSK
Apatity Al
KOVDOR KIROVSK
Kovda-Kuma KANDALAKSHA
VORKUTA
PECHORA COALFIELD
INTA
PECHORA
Kem
Vyg
NADVOITSY ARKHANGELSK
KOMI-UKHTA OIL FIELD
VYARTSILYA Suna
Al UKHTA
PETROZAVODSK PLESETSK
KOTLAS
SYKTYVKAR
Svir
Volkhov
BOKSITOGORSK
NOVGOROD CHEREPOVETS
VOLOGDA
0 Km 300

NORTH-WEST, BALTIC and BELORUSSIA

FINLAND
GULF OF FINLAND
Leningrad
Tallin
Vologda
Volkhov
Narva
Novgorod
BALTIC SEA
Riga
Pskov
Dvina
CENTRE REGION
Neman
Kaunas
Vilnyus
Vitebsk
Kaliningrad
Minsk
Mogilev
Grodno
POLAND
Gomel
Brest
Pripyat
Dnepr
UKRAINE
0 Km 100

PETROZAVODSK
Svir
LENINGRAD
Volkhov
Al CHEREPOVETS
KOLPINO
Narva KIRISHI BOKSITOGORSK
TALLIN
SLANTSY
NOVGOROD
Rybinsk reservoir
PSKOV
VENTSPILS
LIYEPAYA
RIGA
PLAVINAS
KLAYPEDA
KAUNAS
NOVOPOLOTSK
KALININGRAD VITEBSK
VILNYUS ORSHA
KHODINO
GRODNO MOGILEV
MINSK
SOLIGORSK SVETLOGORSK
BEREZA GOMEL
BREST RECHITSA
MOZYR
0 Km 100

0 Km 2000

C Leningrad, Baltic, Belorussia Towns over 50,000

D Leningrad, Baltic, Belorussia Energy, Minerals, Metallurgy

42 THE CENTRE, VOLGA-VYATKA AND BLACK EARTH CENTRE REGIONS

In the system of economic regions in use before the 1960s, these three formed a single region – the Centre – which, as its name implies, played a major role in the economy and organisation of the Soviet state. Covering little more than four per cent of the territory of the USSR, though equivalent in size (916,000 sq. km) to France, West Germany, the Low Countries and Switzerland combined, the area depicted in Map 42 has a population of 45 million, 17 per cent of the Soviet total. Its division into three parts reflects the considerable variety of conditions found within the area. While the present Centre (sometimes referred to as the Industrial Centre) is one of the most highly urbanised (78 per cent) regions of the USSR, the Black Earth Centre, with its *chernozem* soils, is more rural and only 52 per cent of its population live in towns. The Volga-Vyatka region has little unity – the section around Gorkiy is highly industralised and forms a continuation of the large industrial zone between the Oka and Volga rivers, while its more easterly parts (e.g. Kirov oblast) are relatively thinly settled and much less developed.

During the 1970s, the towns of the three regions have continued to grow – though in many cases more slowly than those in other parts of the Soviet Union – while their rural populations have declined by more than 20 per cent. The natural increase rate is very low and there is a migration loss to other regions. Consequently, the total population of the Centre has risen by less than five per cent and those of the Volga-Vyatka and Black Earth Centre have actually declined.

The area as a whole is responsible for rather more than a quarter of total Soviet industrial output – 20 per cent in the Centre proper, which has a higher per capita output than any other region. This leading position in the Soviet economy has been achieved despite the area's relative poverty in industrial resources. The energy base is particularly weak: there is no coal, oil or natural gas and the hydro-electric potential is small; the main resource is the large deposits of lignite and peat which have been extensively used in the generation of electricity. Until recently, the area also appeared to be very poor in other industrial resources with little to offer than a few small iron deposits and some mineral salts. Consequently, the central regions have traditionally relied on coal, pig iron and other raw materials brought in from other regions. Over the past 25 years, the position has changed dramatically: oil and gas pipelines from as far afield as West Siberia and Central Asia give an assured supply of energy and the development of the Kursk Magnetic Anomaly (see p. 48) has made the Black Earth Centre a leading iron ore producer.

These recent trends, however, post-date the establishment within the region of a large industrial capacity. As already indicated (p. 68), the Centre has a major textile industry, established in the nineteenth century, a considerable steel output (p. 56) and a large and growing share of the Soviet Union's chemical industries (p. 66) and is a leading producer of consumer goods of all kinds. The region owes its great importance, as its name suggests, to its centrality with respect to the more densely settled and highly developed European section of the USSR and to its historical function as the organising centre of the Russian Empire and the Soviet Union. It was this region which became the political focus of the Empire in the sixteenth century and its centrality was confirmed, after the Revolution, by the choice of Moscow as the capital and by the establishment there of the most important organs of the highly centralised Soviet state.

The centrality of the region is most clearly expressed in the pattern of communications, both old and new, the great majority of which tend to focus on Moscow. At a time when movement across the European plain was largely by water, the region had easy access to the main arteries, particularly the Volga, and it contains some of the earliest canals built to link the various rivers. When, in the late nineteenth century, rivers and canals were largely replaced by railways, these took a radial form, centred on Moscow, which thus gained direct links with all parts of the Empire. Similar developments are now occurring with respect to the modern road system – it is the roads radiating from Moscow which have been the first to be improved. Again, the rapidly-developing network of internal airlines focuses on Moscow and provides a means of rapid contact with all parts of the Union, still newer forms of 'transport' – electric power lines and oil and gas pipelines – though they cannot perhaps be said to centre on Moscow, feed into the region the power supplies necessary to maintain its varied and growing industries. Thus Moscow sits at the centre of a spider's web of communications and the Centre as a region benefits enormously.

By far the greater part of the area's industrial capacity lies in the Centre proper, particularly in the Moscow-Gorkiy sector between the upper Volga and the Oka river. In the 1960s and 1970s, the Black Earth Centre, formerly an industrial backwater, has been the scene of major industrial developments based on the KMA iron ores which support large new steel plant at Lipetsk and Staryy Oskol as well as supplying ore to other regions. Elsewhere, the aspect is predominantly rural, though the larger towns – especially the oblast capitals – often have important alimentary, engineering and chemical industries.

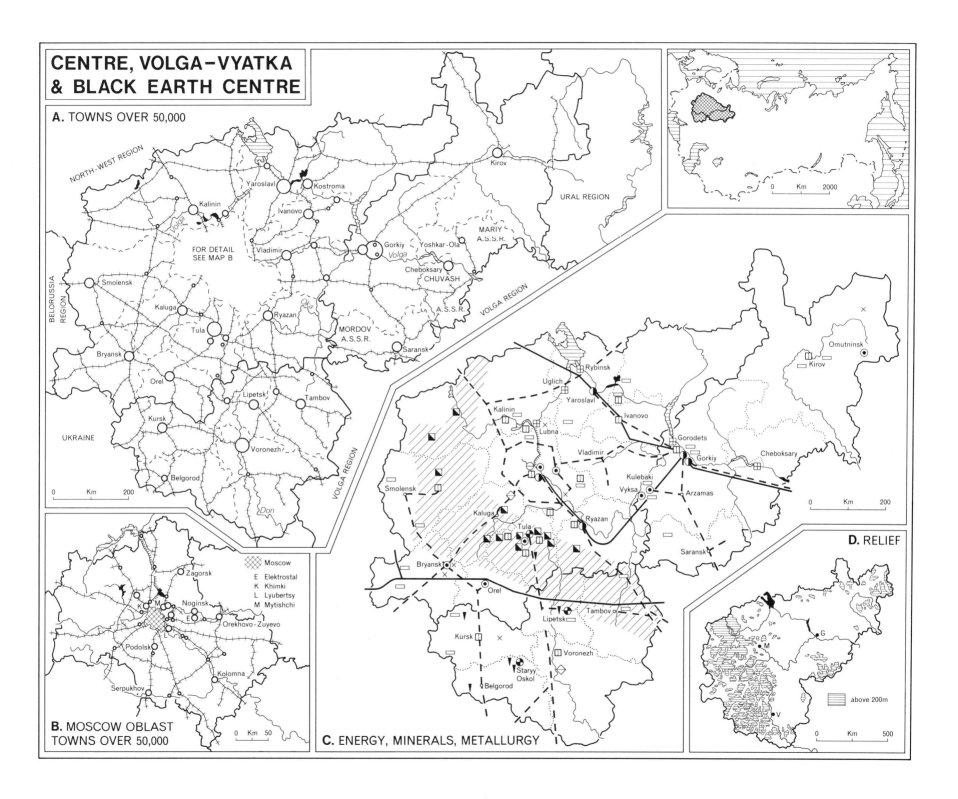

CENTRE, VOLGA–VYATKA & BLACK EARTH CENTRE

A. TOWNS OVER 50,000

NORTH-WEST REGION

Kirov

URAL REGION

Yaroslavl
Kostroma
Kalinin
Ivanovo
Volga
Vladimir
FOR DETAIL
SEE MAP B
Gorkiy
Volga
Yoshkar-Ola
MARIY
A.S.S.R.
Cheboksary
CHUVASH
A.S.S.R.
Smolensk
Oka
Kaluga
Ryazan
MORDOV
A.S.S.R.
Tula
Saransk
BELORUSSIA REGION
Bryansk
Orel
Lipetsk
Tambov
VOLGA REGION
UKRAINE
Kursk
Voronezh
VOLGA REGION
Belgorod
Don

0 Km 200

B. MOSCOW OBLAST
TOWNS OVER 50,000

Zagorsk
Moscow
E Elektrostal
K Khimki
L Lyubertsy
M Mytishchi
Noginsk
Orekhovo-Zuyevo
Podolsk
Kolomna
Serpukhov

0 Km 50

C. ENERGY, MINERALS, METALLURGY

Omutninsk
Kirov
Rybinsk
Uglich
Yaroslavl
Kalinin
Lubna
Ivanovo
Gorodets
Vladimir
Gorkiy
Kulebaki
Cheboksary
Vyksa
Arzamas
Smolensk
Kaluga
Ryazan
Tula
Saransk
Bryansk
Orel
Tambov
Lipetsk
Kursk
Voronezh
Staryy Oskol
Belgorod

0 Km 200

D. RELIEF

G
M
V
above 200m

0 Km 500

0 Km 2000

43 THE UKRAINE AND MOLDAVIA

The area covered by Map 43 includes the Donets-Dnepr, South-western and Southern economic regions of the Ukraine, together with the adjacent small Moldavian republic. Although these territories account for barely three per cent of the land area of the USSR, their combined population of 53.7 million represents 19.5 per cent of the Soviet total and they form one of the most densely settled sections of the country with an overall population density of 85 per sq. km. While the Donets-Dnepr region is one of the most highly urbanised sections of the USSR, with 75 per cent of its population living in towns, the other three regions have a level of urbanisation (50 per cent in 1979) well below the national average. During the 1970s, rural populations in much of the Ukraine have begun to decline, but rural densities in most areas remain high, second only to those of the irrigated farming districts of Transcaucasia and Central Asia.

The high population densities, both rural and urban, in Ukraine/Moldavia reflect the outstanding importance of this zone to the Soviet economy in both the agricultural and industrial sectors, despite recent major developments in other parts of the country. The north-western section of the Ukraine lies in the deciduous forest zone, but the bulk of the territory is characterized by wooded steppe and steppe developed on fertile *chernozem* soils which support highly productive mixed farming systems. 58 per cent of the land area is under crops (compared with 9 per cent for the USSR as a whole) and the region is responsible for about a quarter, by value, of total Soviet agricultural production. Its leading role in agriculture dates from the early nineteenth century when the steppe was colonised, an event accompanied by a massive southward movement of population from the forest zone. The possibilities for further agricultural expansion in the Ukraine are now very limited, save in the drier south-eastern sections where large-scale extension of the irrigated area (Map 43C) is planned.

In the industrial sector, by far the most important section is the easterly Donets-Dnepr economic region, which contains the Donbass coalfield and the Krivoy Rog iron ore deposit supporting the Soviet Union's 'first metallurgical base'. The general development of this leading heavy industrial zone has already been described in the texts accompanying Maps 22 (Minerals), 23 (Coal) and 24 (Iron and Steel); suffice it to add that, despite major industrial developments in other parts of the country, the Donbass-Dnepr bend area still produces nearly a third of the Soviet Union's coal, half its iron ore and more than a third of its crude steel and plays a major role in the heavier branches of engineering.

In addition to its basic resources of coal and iron, the eastern Ukraine has a variety of other industrial resources, notably the manganese deposits at Nikopol and Tokmak, mineral salts at Artemovsk in the Donabass, a major source of natural gas at Shebelinka, south of Kharkov and small oilfields in the Poltava oblast. The hydro-electric potential of the Dnepr river has been exploited by the construction of a series of large hydro-electric stations with a combined capacity in excess of 3 million kW.

The major zone of heavy industry lies in the easternmost Voroshilovgrad and Donetsk oblasts (Maps 43B and D) which have a combined population of 7.9 million, 6.9 million (87 per cent) of whom live in towns. However, large scale urban/industrial development is not confined to the coalfield. The interchange of coal and iron has produced steel and heavy engineering plant at Krivog Rog, at several major cities along the Dnepr (Dnepropetrovsk, Dneprodzerzhinsk, Zaporozhye) as well as at Zhadanov, on the Azov coast, where iron ore is landed from Kerch in the Crimea.

Outside the Donets-Dnepr heavy industrial zone, the economy prior to World War II was predominantly agricultural, though alimentary industries had been established in many of the smaller towns and there were important engineering activities in the major cities of Kiev, Kharkov and Odessa. Apart from the oil, gas and mineral salts of the Lvov and Ivano-Frankovsk oblasts (annexed from Poland at the end of the war), there were few industrial resources in the west. In the post-war period, however, and particularly during the 1960s and 1970s, the western Ukraine and Moldavia have experienced rapid urban growth and accelerated industrialisation. The traditional alimentary industries have expanded and new industries, mainly in the engineering and chemical sectors, have been introduced, drawing on the surplus labour readily available from the densely populated rural districts, trends assisted by the network of oil and gas pipelines built in the region since the 1950s.

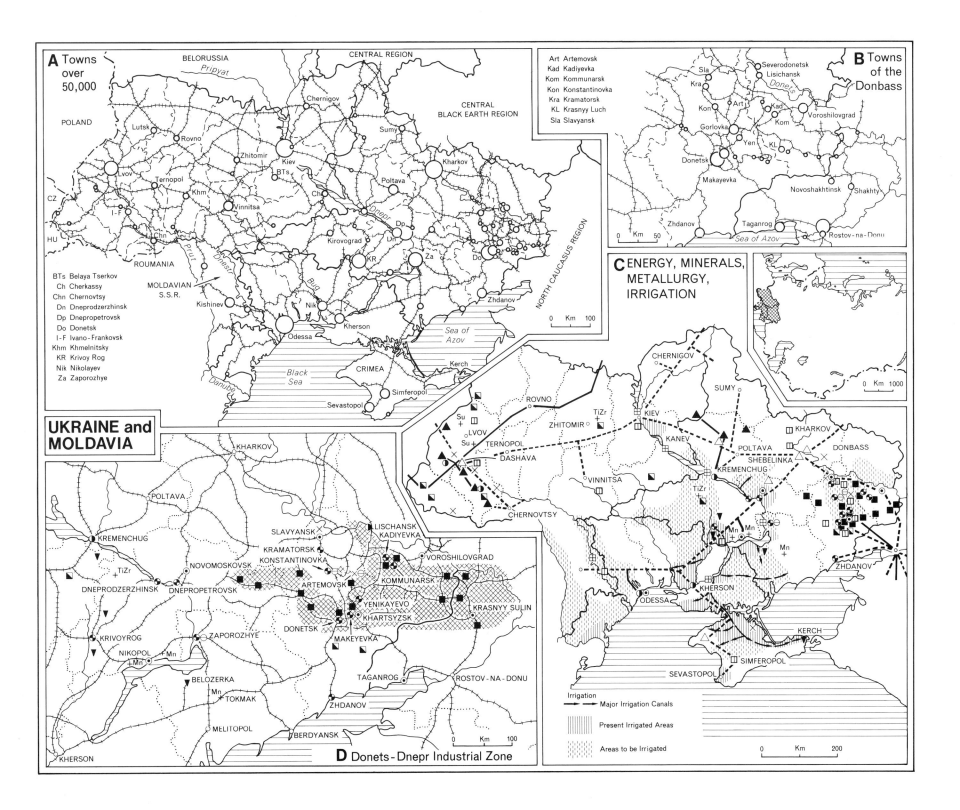

44 THE VOLGA REGION

The Volga economic region comprises a belt of territory some 300–450 km wide strung out along the middle and lower courses of the river for a distance of approximately 1750 km from just above Kazan to its exit into the Caspian. In the early 1960s, the Bashkir ASSR to the north-east, which is topographically part of the Urals, was added to the Volga region, primarily to ensure the operation of the main producing districts of the Volga-Ural oilfield within a single economic area. This gave the region its current extent of 680,000 sq. km (3 per cent of the territory of the USSR); its 1979 population of 19.4 million is just over 5 per cent of the Soviet total.

In its course from north to south, the Volga flows through 10 degrees of latitude (56°N–46°N) and traverses a series of environmental zones from mixed forest to desert (Map E). Consequently there is a long tradition of trade along the river exchanging the varied commodities of these zones. The function of the Volga as a major transport artery, however, involves inter-regional as well as intra-regional exchanges. Waterways, either natural or man-made, connect the Volga north-eastwards to the west flank of the Urals, north-westwards to the Centre and eventually to the White and Baltic seas, south-westwards to the Black Sea and south-eastwards, via the Caspian, to the Transcaucasus and Central Asia. At the same time, the Volga lies athwart major overland routes from the west to the Urals, Siberia, Kazakhstan and Central Asia.

Not surprisingly, towns at the junctions of these land and water routeways became major cities at an early stage, functioning as transshipment points, agricultural market centres and industrial nodes, processing not only the agricultural products of the Volga region itself but also commodities from further afield, such as timber from the forest zone, fish from the Caspian and oil from the Caucasus. The five main centres, spaced evenly at intervals of about 350 km along the river – Kazan, Kuybyshev (formerly Samara), Saratov, Volgograd (formerly Stalingrad and, before that, Tsaritsyn) and Astrakhan – are

ancient cities, all of which had populations well in excess of 100,000 by World War I and had (with the exception of Astrakhan) passed the 350,000 mark by World War II.

Despite these developments, the region remained something of a backwater, at least as far as industry was concerned, until the 1950s and was noteworthy mainly for its agricultural production. In the past quarter-century, the economic significance of the Volga region has changed dramatically, primarily as a result of the vigorous exploitation of its energy resources, both petroleum and hydro-electric.

The development of the Volga-Ural oilfield has already been described (p. 52); a large section of the field lies within the Volga economic region, where the most intensive exploitation has occurred in the northern section, in the Tatar and Bashkir ASSRs and Kuybyshev oblast. In the mid-1960s, when the Volga-Ural field was still the leading producer, these three areas alone produced more than 150 million tonnes of oil a year, 63 per cent of the Soviet total (the Volga region as a whole produced 66 per cent); in the 1970s output has begun to decline but still represents 25 per cent of a very much larger total. Some of the earliest developments of natural gas also occurred in this region, mainly in the Saratov oblast, but the current annual output of 12,000 million cu m is barely 3 per cent of the USSR's natural gas production.

Prior to World War II, the region's large hydro-electric potential remained untouched, but the post-war years have seen the construction of a series of major hydro-electric stations, some of which were in their time the biggest in the country but have since been surpassed by the Siberian stations. The hydro-electric stations at Zhigulevsk (2.3 mill kW, 1955), Volzhskiy (2.5 mill kW, 1961), Balakovo (1.4 mill kW, 1967) and Nizhnekamsk (1.3 mill kW, 1979) together with half a dozen smaller plant which lie in the Ural, Volga-Vyatka and Central regions but are part of the Volga-Kama power system, have a total capacity of 11.25 mill kW – a quarter of the entire hydro-electric capacity of the USSR. In addition there are numerous large thermal stations based mainly on oil and gas, and the

Volga region altogether produces 10 per cent of all Soviet electricity and has a per capita output exceeded only in Siberia. Much of this energy is transmitted, by pipeline and cable, to other parts of the country, but its presence has also permitted the build-up of industry in the Volga region.

Apart from the products of agriculture, there are few non-energy industrial resources. Copper and iron ores are mined in the eastern part of the Bashkir republic, where there are ferrous and non-ferrous metallurgical industries, but this zone is really part of the Urals; elsewhere the only significant items are salt in the Belaya valley, sulphur and some oil shale near Kuybyshev and the mineral salts of Lakes Elton and Baskunchak in the arid south. (The latter produces 4 million tonnes of common salt annually – a quarter of the Soviet supply.)

The traditional alimentary industries, based on local produce, and chemicals, derived from Caucasian oil shipped along the river, have greatly expanded – there are now major oil refining/chemicals complexes in the Bashkir and Tatar republics and around Kuybyshev. In addition there have been major engineering developments, not only in the production of equipment for the oil and electricity industries but also in other fields, notably automobile production at Tolyatti, agricultural machinery at Volgograd (where there is also a small steel mill) and numerous other items. Non-ferrous metallurgy, attracted by the abundant electric power, is carried on at several sites.

Much of the region's industry occurs in the five ancient cities whose populations now range from 460,000 (Astrakhan) to 1.2 million (Kuybyshev) but secondary centres are now growing rapidly. Some of these are satellites of the 'big five', like Tolyatti (502,000) near Kuybyshev and Volzhskiy (209,000) near Volgograd; others form new industrial nodes, as at Ulyanovsk (464,000), Syrzan (179,000) or Naberezhnye Chelny on the lower Kama which had 16,000 inhabitants in 1970 and 301,000 in 1979, based on a major new chemical and engineering complex, which includes the new Kama truck plant.

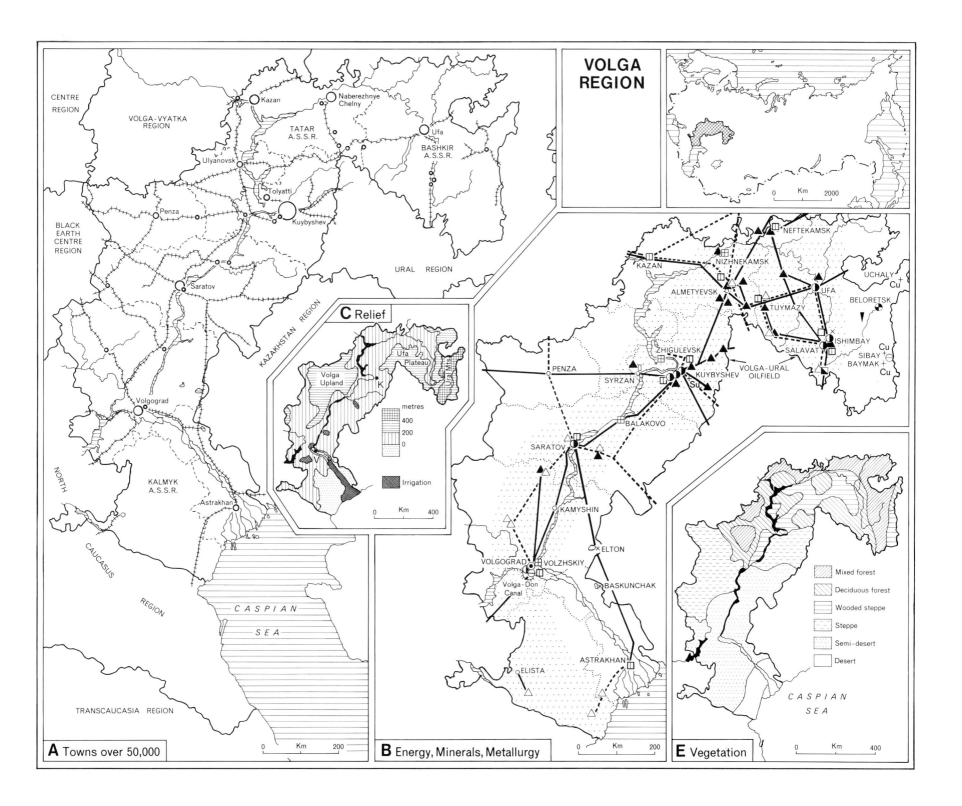

45 THE URAL REGION

Within its current boundaries, the Ural region covers an area of 680,400 sq. km (3 per cent of Soviet territory) and has a population of 15.6 million (6 per cent of the Soviet total); a modest population density (23 per sq. km) and a high degree of urbanisation (74.4 per cent) suggest the presence of large areas of thinly settled territory and the predominance of the industrial sector in the regional economy.

The region's peculiar shape results from the transfer of the Bashkir ASSR to the Volga region, despite which the Ural region still contains a sizeable section of the Volga-Ural oilfield in the Perm and Orenburg oblasts (Map B). A more logical economic boundary would be along the western hill-foot (e.g. the current eastern boundary of the Perm oblast) since the resource bases and economic structures of the zones on either side of such a line are completely different.

Above all else, the Ural region is noteworthy for its metallurgical activities and its role over the past 50 years as the country's 'second metallurgical base' (see p. 56), which is still responsible for nearly a third of Soviet steel output. This role was founded on the region's wealth in iron ore and, despite declining ore production – the Urals now produce only a quarter of Soviet iron ore compared with 40 per cent in 1950 – steel output continues to expand. 23 steel-producing centres are shown on Map D, of which 10 have fully integrated plants and five – Magnitogorsk, Nizhniy Tagil, Chelyabinsk, Novotroitsk and Serov – are responsible for nearly half the region's output, which is well in excess of that of any country in the world outside the USA and Japan despite its poverty in coking coal and its growing iron ore deficit. The development of its links with West Siberia and Kazakhstan have already been described (p. 56).

The Urals are also renowned as a source of ferro-alloy and non-ferrous metal ores (Map C), producing manganese, nickel, chrome, cobalt, platinum, aluminium ores, copper and gold as well as iron. These metallurgical activities are almost all carried on in a central zone some 200 km across comprising the Ural Mountains proper and the hill-foot zone to the east, where it is possible to distinguish half a dozen major industrial complexes centred on the cities of Serov, Nizhniy Tagil, Sverdlovsk, Chelyabinsk, Magnitogorsk and Orsk. In each of these, the mining of metallic minerals, metallurgy and engineering are the predominant activities.

The *Serov* complex in the north has ironworks dating back to the nineteenth century and an integrated steel plant of a million tonnes capacity. The several iron ore sources in the vicinity are supplemented by ores from Kachkanar and the Polunochnoye manganese deposit supports a ferro-alloy plant. In the non-ferrous sector, the Severouralsk bauxite deposit is one of the largest in the Soviet Union and alumina and aluminium are both produced at Krasnoturinsk. Platinum and gold are mined in the mountains to the west. Local energy resources are confined to lignite; coking coal comes from the Kuzbass, natural gas from West Siberia and Central Asia.

The *Nizhniy Tagil* complex is one of the most important steel districts in the USSR with an annual output of 5 million tonnes; Nizhniy Tagil itself produces over 3 million tonnes and there are four other integrated plants and two steel mills (Map D). Based originally on the Nizhniy Tagil magnetite deposit, this district relies increasingly on Kachkanar for iron ore and there are secondary sources near Alapayevsk. Copper, the chief non-ferrous metal, is smelted at Krasnouralsk.

The *Sverdlovsk* complex has no integrated iron and steel plant but produces about 4 million tonnes of steel annually at half a dozen sites which draw their pig iron from Nizhniy Tagil or Magnitogorsk. There is copper smelting at Kirovgrad and at Rezh, which also deals with local nickel ores, and bauxite from Severouralsk is converted to alumina at Sukhoy Log, north of Kamensk-Uralskiy. Chemical and engineering industries are particularly well developed in this zone.

Chelyabinsk, together with a cluster of towns in the northern part of its oblast, forms a fourth major industrial complex. Chelyabinsk itself has an integrated iron and steel plant producing about 4 million tonnes and there are several smaller steel towns to the west. Iron ore came originally from Bakal, but there is growing reliance on the Kustanay deposits in north-west Kazakhstan. Copper is also important and is mined and smelted at Karabash and refined at Kyshtym. Engineering industries are well developed in the western towns, which produce machine tools, railway rolling stock and motor vehicles.

Magnitogorsk, in contrast to the industrial complexes further north, is an isolated single unit with a steel output of some 15 million tonnes from its large integrated works, and supplies crude steel to many parts of the Urals. Although now dependent on Kazakhstan for both ore and coking coal, it remains the biggest single steel producer in the whole of the USSR.

The *Orsk* complex, in the extreme south, while also important for steel, is concerned mainly with non-ferrous metals, chiefly copper and chrome, and has an oil refinery and chemical industries based on oil from the Emba and Mangyshlak fields.

The two industrial complexes in the western section of the Ural region are very different from those listed so far and have developed mainly since World War II. In the Perm oblast, the Solikamsk-Berezniki district has had modern chemical industries since the 1870s which have been much expanded in recent years. From Perm southwards is an energy surplus area producing oil and hydro-electric power which are transmitted to other regions. Finally, the western part of the Orenburg oblast is tied in with the rest of the Volga-Ural oilfield; a new major source of gas has recently been opened up near Orenburg itself and now produces about 10 per cent of Soviet natural gas.

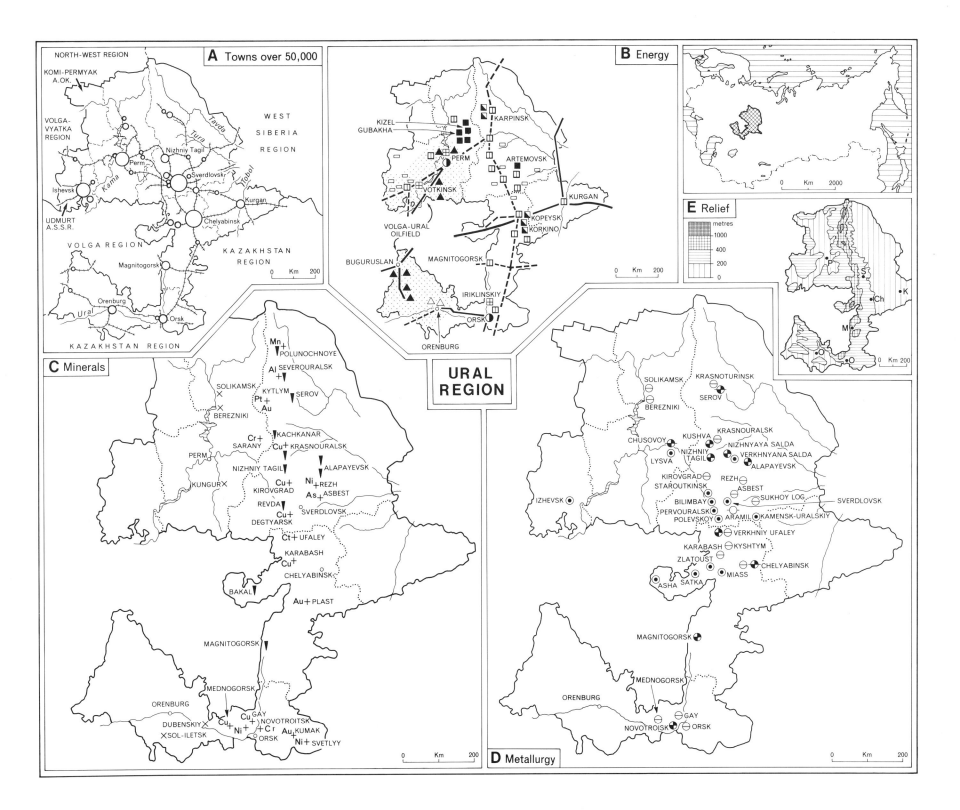

A Towns over 50,000

NORTH-WEST REGION

KOMI-PERMYAK A.OK.

VOLGA-VYATKA REGION

WEST SIBERIA REGION

Tavda
Tura
Tobol
Kama
Nizhniy Tagil
Perm
Sverdlovsk
Ishevsk
Kurgan
Chelyabinsk

UDMURT A.S.S.R.

VOLGA REGION

KAZAKHSTAN REGION

Magnitogorsk

Orenburg
Ural
Orsk

KAZAKHSTAN REGION

0 Km 200

B Energy

KIZEL
GUBAKHA
KARPINSK
PERM
ARTEMOVSK
VOTKINSK
KURGAN
KOPEYSK
KORKINO

VOLGA-URAL OILFIELD

BUGURUSLAN
MAGNITOGORSK
IRIKLINSKIY
ORSK
ORENBURG

0 Km 200

E Relief

metres
1000
400
200
0

P
S
Ch
K
M
O

0 Km 200

URAL REGION

C Minerals

Mn POLUNOCHNOYE
Al SEVEROURALSK
SOLIKAMSK
KYTLYM
Pt SEROV
Au
BEREZNIKI
Cr KACHKANAR
SARANY Cu KRASNOURALSK
PERM
NIZHNIY TAGIL Ni ALAPAYEVSK
REZH
KUNGUR Cu As ASBEST
KIROVGRAD
REVDA SVERDLOVSK
Cu
DEGTYARSK
Ct UFALEY
KARABASH
Cu
CHELYABINSK
BAKAL
Au PLAST

MAGNITOGORSK

MEDNOGORSK
ORENBURG
Cu GAY
DUBENSKIY Cu Ni NOVOTROITSK
SOL-ILETSK Cr KUMAK
ORSK Au
Ni SVETLYY

0 Km 200

D Metallurgy

SOLIKAMSK KRASNOTURINSK
SEROV
BEREZNIKI
KUSHVA KRASNOURALSK
CHUSOVOY NIZHNYAYA SALDA
NIZHNIY VERKHNYANA SALDA
LYSVA TAGIL ALAPAYEVSK
KIROVGRAD REZH
STAROUTKINSK ASBEST
IZHEVSK BILIMBAY SUKHOY LOG SVERDLOVSK
PERVOURALSK ARAMIL KAMENSK-URALSKIY
POLEVSKOY VERKHNIY UFALEY
KARABASH KYSHTYM
ZLATOUST
ASHA SATKA MIASS CHELYABINSK

MAGNITOGORSK

MEDNOGORSK
ORENBURG GAY
NOVOTROISK ORSK

0 Km 200

97

46 SIBERIA AND THE FAR EAST

With a combined area of 12,766,000 sq. km, some 57 per cent of the territory of the USSR, the West Siberian, East Siberian and Far Eastern regions (often referred to collectively as Siberia) are the homes of barely one in every ten Soviet citizens, 27.9 million people in all. Vast northern areas are very thinly settled or uninhabited and, as Map 46A shows, contain only four sizeable towns – the isolated mining centre of Norilsk (180,000); Yakutsk (152,000), the capital and organising centre of the Yakut ASSR; Magadan (122,000), the chief port and fishing centre on the Sea of Okhotsk; and Petropavlovsk in Kamchatka (215,000).

Although these northern areas have huge un-tapped industiral resources, the great bulk of the population and settled area are in the southern districts of the three regions, which also contain most of the agricultural land (Maps D, E), developed resources (Maps C, D, E) and urban-industrial concentrations (Map A) strung out along the lifeline provided by the Trans-Siberian railway.

Some of the earliest industrial developments in the region were associated with the Trans-Siberian railway, which was constructed between 1892 and 1916. The Cheremkhovo coalfield was opened in the 1890s to supply steam coal for the railway and became the basis of a small industrial complex centred on Irkutsk; the field remains of considerable importance, supplying coal to power stations and industrial sites throughout the East Siberian region. The Cheremkhovo-Irkutsk area, however, is now of minor importance compared with the more westerly Kuzbass (Map C). This field, too, was first developed to supply coal to the Trans-Siberian in the late nineteenth century, but its main period of growth dates from the 1930s. The establishment of the Ural-Kuznetsk combine (see p. 56) and the loss of the Donbass during World War II stimulated a rapid expansion of coal output from the Kuzbass, which has continued during the post-war period. The 1979 output of 148 million tonnes was 20 per cent of the Soviet total and second only to that of the Donbass. Metallurgical and heavy engineering industries have also been built up in the Kuzbass; iron ore, formerly brought in from the Urals, now comes from deposits in the hills to the south of the coalfield and from Zheleznogorsk (Map D).

In addition to these major coalfields, lignite and coal have been mined at numerous sites in southern Siberia over the past 40 or 50 years including the Minusinsk basin on the upper Yenisey, the Bukachacha field in the Chita oblast and in the Maritime Kray and Sakhalin in the Far Eastern region (Map E). The most important development of the post-war period has been the exploitation of the Kansk-Achinsk lignite field; output now exceeds 30 million tonnes a year and most is fed into large thermal electricity stations. Coal production in northern areas (Map A) is on a small scale and entirely for local consumption.

Siberia is also extremely rich in a wide variety of metallic minerals. Apart from the isolated Norilsk mining complex, development so far has occurred mainly in the more accessible southerly districts. Lead and zinc ores are mined in the mountains around the Kuzbass; molybdenum, tungsten, tin, lead and zinc in the Buryat ASSR and Chita oblast and in the Khabarovsk and Maritime krays, and there are numerous non-ferrous metal-lurgical industries. Siberia is also the source of most of the Soviet Union's gold.

A major aim throughout the Soviet period has been to build up population and industry in these eastern regions. In Tsarist times and in the early decades of the Soviet period, labour was firmly directed towards the east, but in the more relaxed situation of the post-Stalin period it has proved difficult to establish and maintain the necessary labour supply. Consequently, recent industrial development in these regions has tended to be restricted to a few sectors of special importance to the Soviet economy as a whole and Siberia's role has increasingly been that of supplying energy and raw materials to the older industrial areas in the west; manufacturing industry is not well represented in these regions.

A particularly striking feature of the post-war period has been the development of Siberian energy resources. The major rivers have a huge hydro-electric potential and this has been tapped by the construction of a series of large hydro-electric stations at Novosibirsk, Krasnoyarsk, Bratsk, Angarsk, Ust-Ilim and Sayan. Siberia now produces nearly 20 per cent of all Soviet electric power and has a higher per capita output than any other region.

Prior to the 1950s, oil production was confined to the small Sakhalin field, which supplied re-fineries at Komsomolsk and Khabarovsk (Map E) and much of the oil consumed in Siberia came from the west, along the Trans-Siberian pipeline. Over the past 25 years, major developments have occurred in the West Siberian lowland (Map B) which has proved to have vast resources of both oil and gas. Output in 1979 reached 283 million tonnes of oil and 123,000 million cu m of gas, 48 and 30 per cent respectively of total Soviet production, and further increases are expected.

Finally, mention should be made of the Baykal-Amur Mainline railway (BAM, Map A; see also p. 76) now under construction, which in addition to its role as a new link to the Far East, will facilitate the exploitation of a variety of mineral resources along its route.

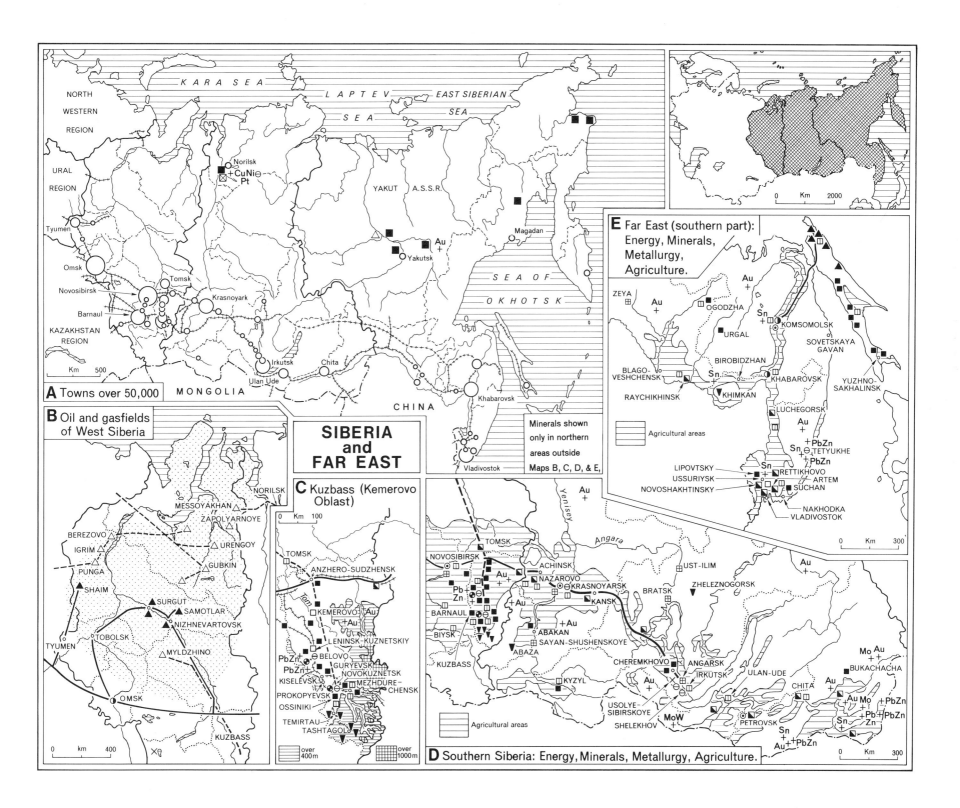

A Towns over 50,000

KARA SEA
LAPTEV SEA
EAST SIBERIAN SEA

NORTH WESTERN REGION

URAL REGION

Norilsk
+CuNi⊖
Pt

YAKUT A.S.S.R.

Magadan

Tyumen

Omsk

Novosibirsk
Tomsk
Krasnoyarsk

Barnaul

KAZAKHSTAN REGION

Km 500

Irkutsk
Chita

Ulan Ude

MONGOLIA

Yakutsk
Au +

SEA OF OKHOTSK

CHINA

Khabarovsk

Vladivostok

Minerals shown only in northern areas outside Maps B, C, D, & E.

SIBERIA and FAR EAST

B Oil and gasfields of West Siberia

NORILSK
MESSOYAKHAN
ZAPOLYARNOYE
BEREZOVO
IGRIM
PUNGA
URENGOY
GUBKIN
SHAIM
SURGUT
SAMOTLAR
NIZHNEVARTOVSK
TYUMEN
TOBOLSK
MYLDZHINO
OMSK
KUZBASS

km 400

C Kuzbass (Kemerovo Oblast)

0 Km 100

TOMSK
ANZHERO-SUDZHENSK
KEMEROVO
Au
Au
LENINSK-KUZNETSKIY
PbZn
BELOVO
PbZn
GURYEVSK
KISELEVSK
NOVOKUZNETSK
PROKOPYEVSK
MEZHDURE-CHENSK
OSSINIKI
TEMIRTAU
TASHTAGOL

over 400m
over 1000m

D Southern Siberia: Energy, Minerals, Metallurgy, Agriculture.

TOMSK
NOVOSIBIRSK
Au +
ACHINSK
NAZAROVO
KRASNOYARSK
UST-ILIM
Au +
Pb Zn
KANSK
BARNAUL
+Au
BRATSK
ZHELEZNOGORSK
BIYSK
+Au
ABAKAN
SAYAN-SHUSHENSKOYE
KUZBASS
ABAZA
CHEREMKHOVO
ANGARSK
ULAN-UDE
Au +
KYZYL
IRKUTSK
Mo Au
+ +
BUKACHACHA
USOLYE-SIBIRSKOYE
SHELEKHOV
MoW
PETROVSK
CHITA
Au
+
Au Mo
+ +
PbZn
Sn
+ +
Au +PbZn
Pb +PbZn
Zn

Yenisey
Au +
Angara

Agricultural areas

0 Km 300

E Far East (southern part): Energy, Minerals, Metallurgy, Agriculture.

ZEYA
Au
Au +
OGODZHA
URGAL
Sn +
KOMSOMOLSK
SOVETSKAYA GAVAN
BLAGO-VESHCHENSK
Sn
BIROBIDZHAN
KHABAROVSK
YUZHNO-SAKHALINSK
RAYCHIKHINSK
KHIMKAN
LUCHEGORSK
Au +
+PbZn
TETYUKHE
Sn +
⊖+PbZn
LIPOVTSKY
USSURIYSK
NOVOSHAKHTINSKY
Sn +
RETTIKHOVO
ARTEM
SUCHAN
NAKHODKA
VLADIVOSTOK

Agricultural areas

0 Km 300

0 Km 2000

47 THE NORTH CAUCASUS AND TRANSCAUCASIA

The North Caucasus region of the RSFSR, together with the three Transcaucasian republics – Georgia, Armenia and Azerbaydzhan – occupy the isthmus between the Black and Caspian seas and form a highly distinctive region sometimes referred to as Caucasia. While their combined area of 541,000 sq. km is only 2.4 per cent of the territory of the USSR, they have a population of 29.6 million, 11.3 per cent of the Soviet total. Their overall population density of 55 per sq. km is about five times the Soviet average, suggesting a zone of quite intensive economic activity, but this figure masks a great deal of local variation imparted by the region's strong relief (Map E). Densities are low throughout the extensive mountain areas and high in the hill-foot zone on the north side of the main Caucasus range and in the Transcaucasian lowlands of the Rioni and Kura basins.

These variations owe much to the predominance of agriculture in the regional economy: 45 per cent of the population is still classed as rural and there was relatively little industry before World War II. As Map D indicates, the northern part of the region is dominated by the cereal and livestock farming characteristic of the European steppe (see p. 44), with some, more intensive, irrigated agriculture in the Don and Kuban valleys. The hillfoot zone of the main Caucasus ranges and the Transcaucasian lowlands, however, specialise in the production of high value crops, often grown under irrigation, such as maize, rice, oilseeds, fruits (both pome and citrus), vines, tea, tung nuts, tobacco and cotton, as well as wheat and sugar beet. This agricultural wealth underlies much of the region's traditional prosperity, particularly that of Georgia and Armenia, which are among the richer Soviet republics. Not surprisingly, alimentary industries are well developed.

Caucasia is also rich in industrial raw materials, particularly energy and metallic minerals. The inclusion of the northernmost, Rostov, oblast in the North Caucasus region of the RSFSR appears illogical in view of its environmental and economic similarities to the adjacent eastern Ukraine. The section north of the Don is in fact a continuation of the Donbass coalfield metallurgical zone and produces about 30 million tonnes of coal annually (one seventh of the Donbass total), which supports several large thermal generating plant (Map B) and the steel mills at Krasnyy Sulin and Taganrog (Map C). Elsewhere there are only minor coal resources. The most important fields are at Tkvarcheli and Tkibuli in Georgia; these produce about 2 million tonnes a year, most of which is coking coal fed to the 1.5 million tonne capacity integrated steelworks at Rustavi, which draws its iron ore from Dashkesan in western Azerbaydzhan. Georgia also has a steel rolling mill and ferro-alloy plant at Zestafoni, the latter dependent on the major manganese deposit at Chiatura. There is also a small steel mill, producing pipes for the oil industry, at Sumgait, near Baku.

The region is most renowned for its oil and gas resources, though these are much less significant than was once the case. A large petroleum basin extends across the isthmus from the Sea of Azov to the Caspian and oil was extracted around Maykop, Groznyy and, above all, Baku in the nineteenth century; much of the output went by pipeline to Batumi for export. These North Caucasus oilfields produced the bulk of the Soviet oil supply throughout the inter-war period and as late as 1950 were responsible for more than half the total output (21 out of 38 million tonnes). Since then, however, developments elsewhere have forced the Caucasus into a minor position. After reaching a peak output of 55 million tonnes in 1970, achieved largely by increased exploitation of submarine sources at Baku, production has declined: in 1979 it was only about 35 million tonnes, a mere 6 per cent of Soviet production. The region's natural gas resources were developed rapidly in the late 1950s and early 1960s when major deposits were opened up in the Krasnodar and Stavropol krays and around Baku. In the mid-1960s, these areas produced 45,000 million cu m annually (40 per cent of the Soviet total) but, by the late 1970s, despite increased output from Azerbaydzhan, this had fallen to 28,000 million cu m (7 per cent). Nevertheless, the exploitation of oil and gas has provided energy for industrial development in the Caucasus and supports several major petro-chemical complexes.

The strong relief and heavy precipitation occurring in parts of the region gives a considerable hydro-electric potential and numerous plant have been built on the small, swift-flowing rivers. Three larger 'second generation' schemes have also been constructed – the Razdan 'cascade' of six stations (on the Razdan river which descends from Lake Sevan to Yerevan) with a capacity of 1.1 million kW was completed in 1974, the Chirkey station (1.0 mill kW) in Dagestan in 1976 and the Inguri station (1.6 mill kW) in Georgia in 1979.

The availability of electric power and the presence of metallic minerals has led to the development of several non-ferrous metallurgical complexes. The oldest of these is at Ordzhonikidze, on the north flank, where lead and zinc have been produced since the 1860s, but most developments have been post-World War II. Copper is mined and smelted in the Madneuli-Alaverdi, Kafan and Kadzharan districts (Map C) and there is a major molybdenum-tungsten source at Tyrnyauz in the Kabardino-Balkar ASSR. Attempts to develop aluminium production based on alunite from Alunitdag and nephelite from Tezhsar to supply aluminium plant at Yerevan and Sumgait have experienced technical problems.

The Caucasus thus has a mixed industrial structure, involving energy production, metallurgy, both ferrous and non-ferrous, and a variety of alimentary chemical and engineering activities.

Note: The ethnic diversity of the region is reflected in its complex politico-administrative structure. The 17 territorial divisions indicated on Map A are:— North Caucasus: Ro = Rostov oblast; Kr = Krasnodar kray (includes Ad = Adygey autonomous oblast); St = Stavropol kray (incl. KC = Karachayevo-cherkess aut. ob.); KB = Kabardino-Balkar ASSR; NO = North Ossetian ASSR; CI = Chechen-Ingush ASSR; Da = Dagestan ASSR; Transcaucasia: Ge = Georgian SSR (incl. Ab = Abkhaz ASSR; Ad = Adzhar ASSR; SO = South Ossetian aut. ob.); Ar = Armenian SSR; Az = Azerbaydzhanian SSR (incl. NK = Nagorno-Karabakh aut. ob.; Na = Nakhichevan ASSR).

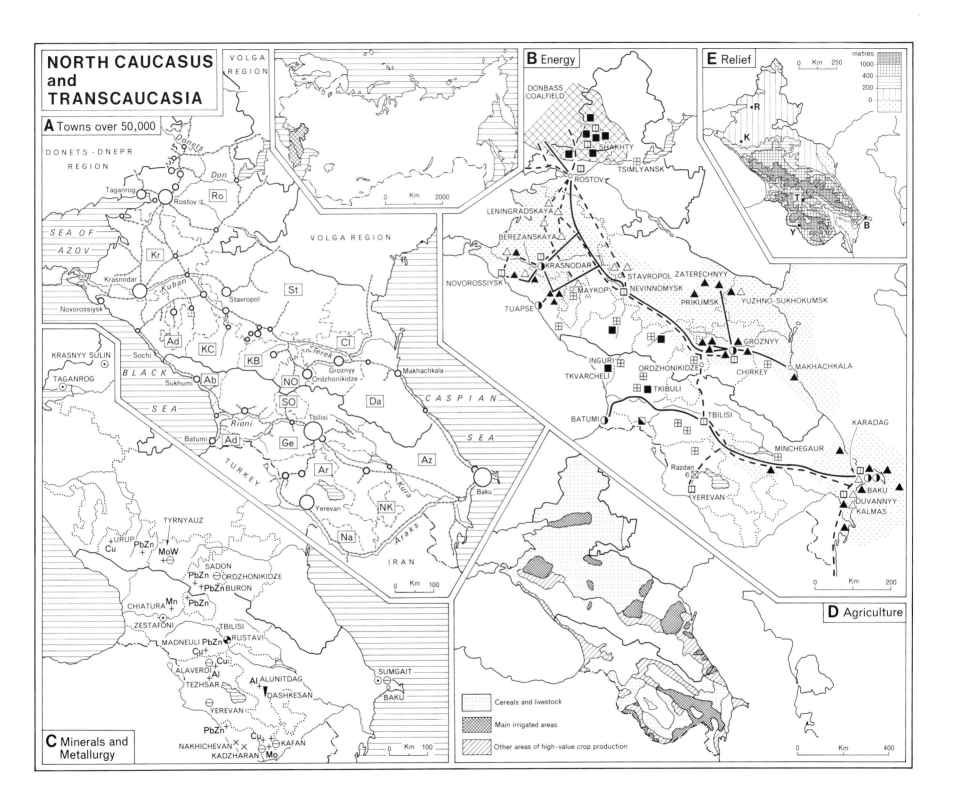

48 KAZAKHSTAN

With an area of 2,717,300 sq. km, Kazakhstan is the second largest of the 15 Union Republics and has a population of 14.7 million. The Kazakh SSR is also classed as a single Major Economic Region, but this is less a reflection of economic reality than of adherence to the idea of an ethnically homogenous political entity with some economic autonomy. Even the notion of ethnic homogeneity is unreal; the republic's population is barely one-third Kazakh; 52 per cent are Russians, Ukrainians or Belorussians and the remainder are members of a score of other nationalities.

Diversity rather than uniformity is apparent in the physical environment. The northern section of the republic is steppeland, which continues across the now wholly artificial Kazakh/RSFSR boundary into Western Siberia; both regions were affected by the Virgin Lands scheme (see p. 44), which brought in large numbers of Russian settlers. The central section is arid or semi-arid (note the 350 km canal built to supply water from the Irtysh to the Karaganda industrial complex) and is notable mainly for the mineral wealth of the Kazakh upland in the east. The southern fringes of Kazakhstan form a continuation of the hill-foot zone of Central Asia, while the west comprises a southerly extension of the Ural ranges and part of the lowland around the Caspian. Each of these sections has its own distinctive resource endowment and industrial development has occurred in several widely separated complexes with very little integration.

Kazakhstan plays a significant role in Soviet heavy industry and the most important industrial complex is still that centred on the Karaganda coal and metallurgical area. As indicated earlier (p. 50), the Karaganda coalfield was opened up in the 1930s to supply coking coal to Magnitogorsk and the return freight of iron ore led to the establishment of a steel industry. This has been expanded to a capacity of 6 million tonnes of steel per annum, produced in the modern integrated plant at Temir Tau and a smaller steel mill at Yermak in the Irtysh valley. Ural iron ore is no longer available and the Kazakh steel industry relies on the Atasu and Karazhal deposits in the arid central zone and those of the Kustanay oblast in the north-west, which also supply the Urals.

Some 200 km to the north-east of Karaganda, a more recent development has occurred in the Ekibastuz district which now produces some 60 million tonnes of lignite and low grade bituminous coal; mined open-cast, these are fed into large thermal generating plant nearby.

The oil and gas resources of Kazakhstan are limited. The Guryev (Emba) oilfield in the extreme west has been in operation since the late nineteenth century, but annual output has never exceeded 3 million tonnes. Much was expected of the Mangyshlak field further south, the development of which involved the construction of new settlements, railways and pipelines, together with a nuclear electricity station and a desalinisation plant to provide water from the Caspian, but, after reaching a peak of about 20 million tonnes in 1975, output has declined. In 1979, Kazakhstan produced only 18 million tonnes of oil and 5,000 million cu m of gas, 6 and 1.4 per cent respectively of the Soviet total. While the production of the Emba and Mangyshlak fields is piped out of the republic, oil comes in from West Siberia to the Pavlodar refinery and gas from Central Asia is the main energy source for southern Kazakhstan.

The republic is especially significant in the production of non-ferrous metals, which forms the basis of several important industrial complexes. One of the biggest of these is in the extreme east, in the upper Irtysh valley, which is now the Soviet Union's leading source of lead and zinc and also produces copper and gold. Power comes mainly from hydro-electric plant on the Irtysh. Over Kazakhstan as a whole, copper is the leading metal and the republic now rivals the Ural region in this respect; most of the output is from the Kounradskiy-Balkhash-Sayak and Karsakpay-Dzhezkazgan complexes. Lead, zinc and antimony are mined around Tekeli in the south-east and lead and zinc in the Chimkent oblast, while in the north-west nickel and chrome are mined near Aktyubinsk, where they support a ferro-alloy plant drawing its pig-iron from the Urals. A more recent development has been the exploitation of bauxite at Arkalyk and Krasnooktyabrskiy. The bauxite is fed to an alumina plant at Pavlodar whence alumina is sent to the Siberian aluminium reduction plant.

In addition, Kazakhstan produces a variety of other minerals of which the most important are the phosphates of Karatau, near Dzhambul.

Thus the republic is of outstanding importance to the Soviet economy as a supplier of non-ferrous ores and metals and contributes coking coal and iron to the Ural steel-producing districts. The northern steppe zone is a major source of grain for the European regions.

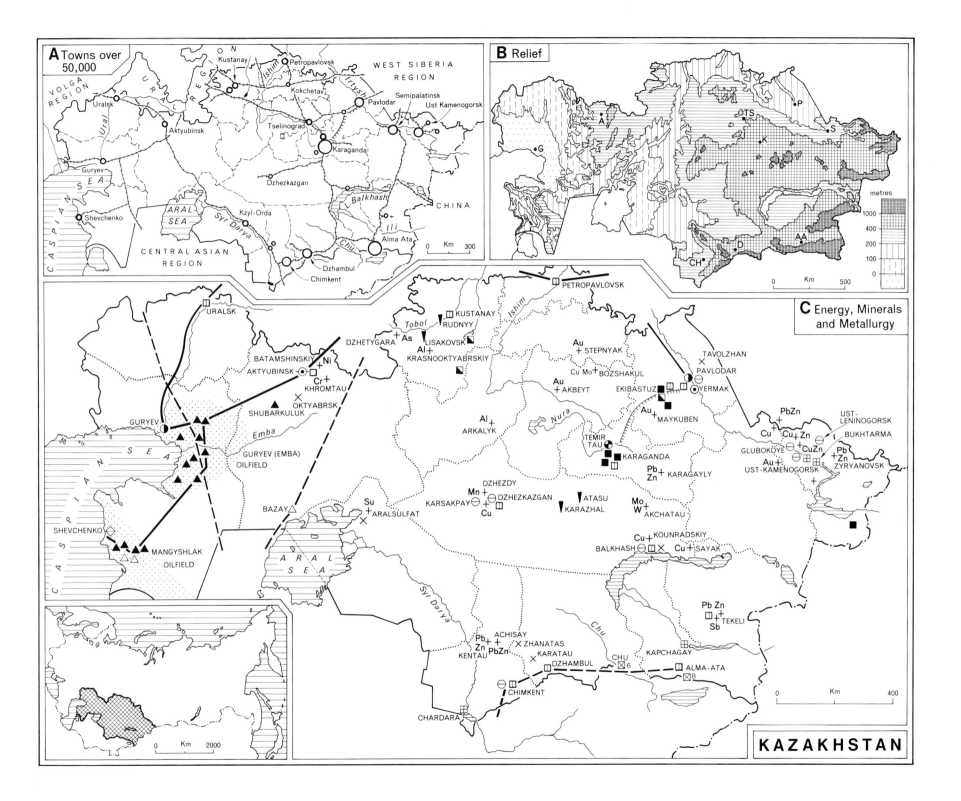

49 CENTRAL ASIA

With a combined area of 1,277,100 sq. km, the four Central Asian republics of Uzbekistan, Kirgiziya, Tadzhikistan and Turkmeniya account for only 5.7 per cent of the territory of the USSR but contained 9.7 per cent of its population – 25,480,000 people – in 1979. Despite recent large-scale immigration, less than 20 per cent of the region's inhabitants are of European origin and Central Asia remains very different, culturally, from the rest of the USSR. Rapid population growth (28.7 per cent 1970–79), both urban and rural, and a low level of urbanisation (41 per cent) are other characteristic features.

The region displays major physical contrasts (Map 49B). The eastern half, covering the Kirgiz and Tadzhik republics and eastern Uzbekistan, is dominated by high mountain ranges with restricted valley and basin lowlands; the greater part of Uzbekistan and almost the whole of the Turkmen republic form part of the vast Central Asian desert plain. This arrangement has had a fundamental effect on the distribution of both population and economic activities which were, until quite recently, very heavily concentrated in the agricultural sector. Most of the inhabitants of Central Asia live in the piedmont zone between the mountains and the desert and along the flood plains of the Syr Darya and Amu Darya, supported by intensive irrigated agriculture (Map 49C). Despite its predominantly agricultural aspect and low level of urbanisation, the region is the site of several extremely ancient cities – Tashkent, Bukhara, Samarkand – which developed as the centres of small Moslem states and trading centres on the caravan routes from China.

Agriculture remains the leading economic activity, supplying high value sub-tropical crops, notably cotton, to other parts of the USSR. The Soviet period has witnessed large-scale investment in new irrigation works but the concentration of effort on the production of high-value crops, together with continuing rapid population growth have resulted in a deficiency in basic foodstuffs, particularly grain, large quantities of which have to be brought in from Kazakhstan.

Prior to the 1950s, industrial development was very limited, a situation due only in part to the fact that the region lacked major sources of coal and iron which elsewhere were the basis of the first stage in the industrialisation of the USSR. Central Asia's natural resources have proved more relevant to the second stage based on oil, gas, electricity, non-ferrous metals and chemicals.

In the extreme west, the Turkmen oilfields were first developed in the late nineteenth century, but the bulk of their output was shipped across the Caspian and had little influence on development in Central Asia, which relied on the small coal and oil deposits in the Fergana basin and elsewhere for its limited energy requirements. Production from the Turkmen fields (now consumed mainly within the region) peaked at about 16 million tonnes, three per cent of the Soviet total, in the early 1970s and is expected to decline. The inter-war and early post-war years saw some development of the region's rich resources of non-ferrous metallic and other minerals and the expansion of manufacturing industry, particularly cotton textiles. A small steel plant was built at Bekabad, using local scrap and pig-iron from Karaganda and the Kuzbass, but its annual output of 400,000 tonnes now represents less than 20 per cent of the region's needs.

The past 25 years have witnessed major developments in several sectors, of which the most outstanding has been natural gas. This was first discovered in the early 1960s around Bukhara in the Amu Darya valley and subsequent exploration has disclosed a major gas field beneath the deserts of Uzbekistan and Turkmeniya. Production from Central Asia as a whole in the late 1970s exceeded 100 billion cu m annually, about a quarter of Soviet output. A regional pipeline from the Amu Darya via Tashkent and Frunze to Alma-Ata (in Kazakhstan) supports large thermal generating plant in several cities. In addition, long-distance pipelines carry large quantities of Central Asian gas to the Urals and the Centre.

In addition to these mineral fuel resources, Central Asia has a large hydro-electric potential though its development conflicts with the needs of irrigation in that the former requires large reservoirs at a constant level, while the latter requires the release of their contents during the summer months. Several small plant supply local needs and a few major ones have been built during the 1960s and 1970s, notably the Chirchik-Bozsu cascade of 16 stations producing 1.2 million kW, Toktogul (1.2 m kW) on the Naryn river and Nurek (2.7 kW) on the Vakhsh; a 3.6 m kW station is under construction on the latter river.

A variety of modern industrial complexes have been established in recent years, of which the following are examples. Navoi, 150 km east of Gazli, where natural gas was first discovered, is now a major chemical centre, producing nitrogenous fertilisers, cellulose acetate, acrylic fibres and cement. At Angren, in Tashkent oblast, an important lignite deposit is overlain by kaolinite from which it is hoped to supply alumina for conversion to aluminium at Regar, near the Nurek hydro-electric station. Also in Tashkent oblast, Almalyk has copper, lead, zinc, tungsten, and molybdenum refineries and produces ammonium phosphate using phosphatic rock from Karatau and nitrogen from Chirchik. Gaurdak, in the extreme south-east of the Turkmen republic, is a major source of sulphur. Among developments planned for the immediate future, one of the largest is to be at Chardzhou on the Amu Darya, where an oil refinery is to be built and will be supplied with oil by a pipe line from western Turkmeniya. Refinery products will be used in the manufacture of plastics, synthetic fibres and synthetic rubber. Meanwhile the more traditional industries – cotton textiles, light engineering and food processing – continue to expand, mainly in the older urban centres.

Despite the more rapid pace of industrialisation in recent years, population growth has resulted in a marked labour surplus in rural areas, whose inhabitants appear unwilling to move in large numbers to urban areas and are even less likely to shift to unfamiliar parts of the USSR (e.g. Siberia) where there is a labour shortage. Consequently, much of the urban/industrial growth which has occurred in Soviet Central Asia has depended on the immigration of workers from the European regions and much of its raw material wealth – cotton, gas, non-ferrous metals and other minerals – is still 'exported' to other regions of the USSR.

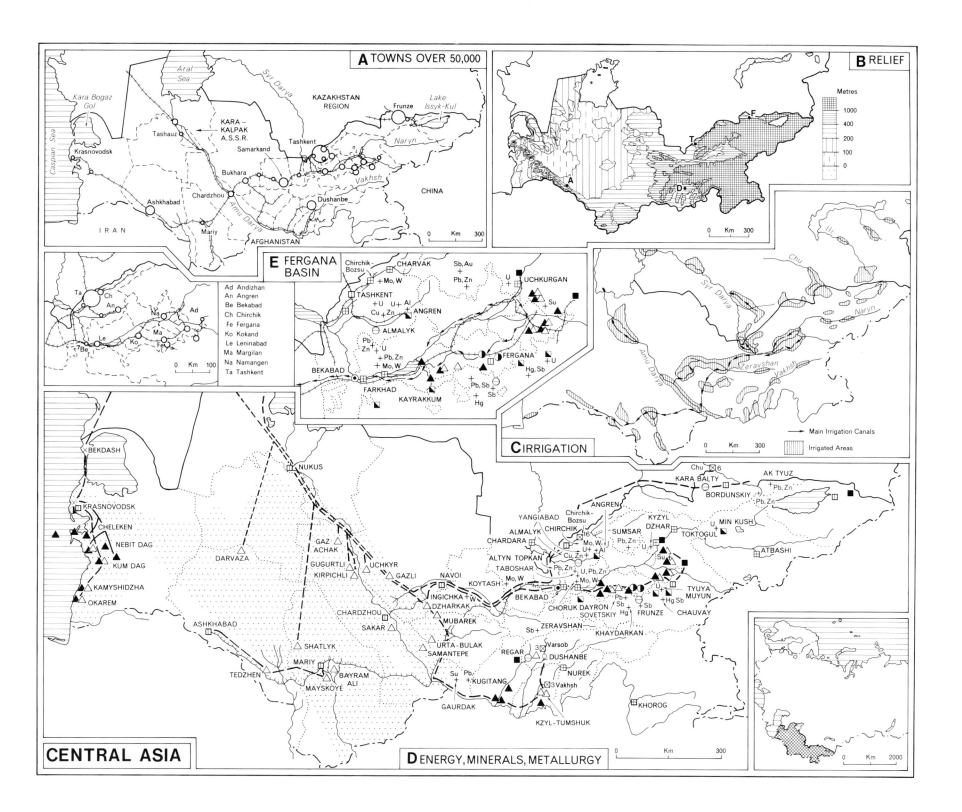

BIBLIOGRAPHY

1. General geographical texts

COLE, J. P., *Geography of the* USSR, Penguin Books, Harmondsworth, 1967.

COLE, J. P. and GERMAN, F. C., *A Geography of the* USSR – *The Background to a Planned Economy*, Butterworths, London, 1961, 2nd edn. 1970.

CRESSEY, G. B., *Soviet Potentials – A Geographical Appraisal*, Syracuse University Press, Syracuse, N. Y. 1962.

DEWDNEY, J. C., *A Geography of the Soviet Union*, Pergamon Press, Oxford, 1965, 2nd edn. 1971, 3rd edn. 1979.

FRENCH, R. A., *The* USSR *and Eastern Europe*, Oxford University Press, London, 1965.

GREGORY, J. S., *Russian Land, Soviet People*, Harrap, London, 1968.

HOOSON, D. J. M., *The Soviet Union – A Systematic Regional Geography*, University of London Press, London, 1966.

HOWE, G. M., *The Soviet Union*, Macdonald and Evans, London, 1968.

HOWE, G. M., *The* USSR, Hulton Educational, Amersham, 1972.

JORRE, G., *The Soviet Union – The Land and its People*, Longmans, London, 1950, 2nd edn. 1961, 3rd edn. 1967 (tr E. D. Laborde, revised by C. A. Halstead).

LYDOLPH, P. E., *Geography of the* USSR, Wiley, New York, 1964, 2nd edn. 1970, 3rd edn. in two volumes: *Geography of the* USSR, Wiley, New York, 1977; *Geography of the* USSR *– Topical Analysis*, Misty Valley Publishing, Elkhart Lake, Wisconsin, 1979.

MELLOR, R. E. H., *Geography of the* USSR, Macmillan, London, 1964.

MINTS, A. A., *The Geography of the* USSR *– An Introductory Survey*, Novosti, Moscow, 1975.

PARKER, W. H., *The World's Landscapes: 3 The Soviet Union*, Longman, London, 1969.

PARKER, W. H., *The Super-Powers: The United States and the Soviet Union Compared*, Macmillan, London, 1972.

POKSHISHEVSKIY, V. V., *The Geography of the Soviet Union*, Progress Publishers, Moscow, 1974.

SYMONS, L. (ed) *The Soviet Union: A Systematic Geography*, Hodder and Stoughton, 1982.

2. Physical environment

BERG, L. S., *The Natural Regions of the* USSR, Macmillan, New York, 1950.

BORISOV, A. A., *Climates of the* USSR, Oliver and Boyd, Edinburgh, 1966 (trans. R. A. Ledward; ed. C. A. Halstead).

FRENCH, R. A., "Conservation and pollution in the USSR", *Geographical Journal*, 139 (3), 1973, 521–524.

GERASIMOV, I. P., ARMAND, D. L., and YEFRON, K. M., *Natural Resources of the Soviet Union: Their Use and Renewal*, Freeman, San Francisco, 1971.

GOLDMAN, M. I., *The Spoils of Progress: Environmental Pollution in the Soviet Union*, MIT Press, Cambridge, Mass., 1972.

LYDOLPH, P. E., *Climates of the Soviet Union* (World Survey of Climatology), Elsevier, Amsterdam, 1977.

MICKLIN, P. P., "Disciplinary plans for Soviet rivers", *Geographical Magazine*, LI(10), July 1979, 701–706.

NALIVKIN, D. V., *The Geology of the* USSR, Pergamon Press, Oxford, 1960.

PRYDE, P. R., *Conservation in the Soviet Union*, Cambridge University Press, London, 1972.

RIKHTER, G., PREOBRAZHENSKIY, V., and NEFEDYEVA, V., "The Soviet land revealed", *Geographical Magazine* XLVIII(5), February 1976, 266–272.

SHTILMARK, F., "Network of nature reserves", *Geographical Magazine*, XLVIII(12), September 1976, 742–746.

SINGLETON, F., *Environmental Misuse in the Soviet Union*, Praeger, New York, 1976.

SUSLOV, S. P. *Physical Geography of Asiatic Russia*, Freeman, San Francisco, 1961.

VELITCHKO, A. A., "Soviet glaciers were late developers", *Geographical Magazine*, LI(7), April 1979, 472–478.

3. Social, cultural and historical

ANDERSON, B. A. "Data sources in Russian and Soviet demography" *in* Kosinski, *op. cit.* 1977.

AZRAEL, J. R. (ed) *Soviet Nationality Policies and Practices*, Praeger, New York, 1978.

BALL, B. and DEMKO, G. J., "Internal migration in the Soviet Union", *Economic Geography*, 54(2), 1978, 95–113.

BATER, J. H., *The Soviet City, Ideal and Reality*, Arnold, London, 1980.

CZAP, P., "Russian history from a demographic perspective" *in* Kosinski, *op. cit.*, 1977.

DAVIES, R. W. (ed) *The Soviet Union*, Allen and Unwin, London, 1978.

DEMKO, G. J. "Demographic research on Russia and the Soviet Union", *in* Kosinski, *op. cit.*, 1977.

DESFOSSES, H., "The USSR and the world population crisis" *in* Kosinski, *op. cit.*, 1977.

DEWDNEY, J. C., "Population changes in the Soviet Union, 1959–1970", *Geography*, 56(4), 1971, 325–330.

DEWDNEY, J. C., "Inquiry into people: the USSR", *Geographical Magazine*, XLVI(7), April 1974, 350–354.

FRENCH, R. A., "The individuality of the Soviet City" *in* French and Hamilton, *op. cit.*, 1979.

FRENCH, R. A. and HAMILTON, F. E. I., "Is there a socialist city?" *in* French and Hamilton, *op. cit.*, 1979.

FRENCH, R. A. and HAMILTON, F. E. I. (eds.), *The Socialist City*, Wiley, Chichester, 1979.

GILBERT, M., *Imperial Russian History Atlas*, Routledge and Kegan Paul, London, 1978.

GILBERT, M., *Soviet History Atlas*, Routledge and Kegan Paul, London, 1979.

HAMILTON, F. E. I., "Modern Moscow planned in rings", *Geographical Magazine*, XLV(6), March 1973, 451–459.

HARRIS, C. D., "City and region in the Soviet Union" in *Urbanization and its Problems* (R. P. Beckinsale and J. M. Houston eds.), Blackwell, Oxford, 1968.

HARRIS, C. D., *Cities of the Soviet Union: Studies in their Functions, Size, Density and Growth*, Rand McNally, Chicago, 1970.

HARRIS, C. D., "Urbanization and population growth in the Soviet Union, 1959–1970", *Geographical Review* LXI(1), 1971, 102–124.

HOOSON, D. J. M., "The growth of cities in pre-Soviet Russia" in *Urbanization and its Problems* (R. P. Beckinsale and J. M. Houston eds.), Blackwell, Oxford, 1968.

KOSINSKI, L. A. (ed.) *Demographic Developments in Eastern Europe*, Praeger, New York, 1977.

KOVALEV, S. "Farewell to the rural scene", *Geographical Magazine*, XLVIII(7), April 1976, 427–432.

KROTKI, K. J., "Fertility and KAP surveys in eastern Europe and the Soviet Union" *in* Kosinski *op. cit.*, 1977.

LEWIS, R. A. and ROWLAND, R. H., "Urbanization in Russia and the USSR, 1897–1966, *Annals Assoc. American Geographers*, 59, 1969, 776–796.

LEWIS, R. A. and ROWLAND, R. H. *Population Redistribution in the* USSR, *Its Impact on Society, 1897–1977* Praeger, New York, 1979.

LEWIS, R. A., ROWLAND, R. H. and CLEM, R. S., *Nationality and Population Change in Russia and the* USSR: *An Evaluation of Census Data, 1897–1970*, Praeger, New York, 1976.

LISTENGURT, F., "Soviets seek the city lights", *Geographical* Magazine, XLVIII(8) May 1976, 492–496.

LORIMER, F., *The Population of the Soviet Union – History and Prospects*, League of Nations, Geneva, 1946.

LYDOLPH, P. E., "Manpower problems in the USSR", *Tijdschrift voor Econ. en Soc. Geografie*, 1972, 331–344.

LYDOLPH, P. E. and PEASE, S. R., "Changing distribution of population and economic activities in the USSR", *Tijdschrift voor Econ. en Soc. Geografie*, 1972, 244–261.

MATTHEWS, M. *Class and Society in Soviet Russia*, Lane/Penguin, London, 1972.

MATTHEWS, M., "Social dimensions in Soviet urban housing" *in* French and Hamilton, *op. cit.* 1979.

MILLER, J., *Life in Russia Today*, Batsford, London, 1969.

NETTL, J. P. *The Soviet Achievement*, Thames and Hudson, London, 1967.

PARKER, W. H., *An Historical Geography of Russia*, University of London Press, London, 1968.

PARKER, W. H., *The Russians: How They Live and Work*, David and Charles, Newton Abbot, 1973.

REINER, T. A. and WILSON, R. H., "Planning and decision-making in the Soviet City" *in* French and Hamilton, *op. cit.*, 1979.

SHABAD, T., "Soviet migration patterns based on the 1970 census data" in Kosinski, op. cit., 1977.

SHAW, D. J. B. "Recreation and the Soviet city", in French and Hamilton, op. cit., 1979.

SMITH, H., The Russians, Times Books, London, 1976.

4. Economics, economic geography (general)

BANDERA, V. N. and MELNYK, Z. L., The Soviet Economy in Regional Perspective, Praeger, New York, 1973.

CAMPBELL, R. W., The Soviet-Type Economies: Performance and Evolution, Houghton Mifflin, Boston, 1974.

CLARKE, R. A., Soviet Economic Facts, 1917–1970, Macmillan, London, 1972.

DEWDNEY, J. C., The USSR (Studies in Industrial Geography), Dawson, Folkestone, 1976/Hutchinson, London, 1978.

DIENES, L., "Investment priorities in Soviet regions", Annals Assoc. Amer. Geographers, 62(3), 1972, 437–454.

DOBB, M., Soviet Economic Development since 1917, Routledge and Kegan Paul, London, 1948, 2nd edn. 1966.

DYKER, D. A., The Soviet Economy, Crosby Lockwood Staples, London, 1976.

HUNTER, H., The Future of the Soviet Economy: 1978–1985, Westview Press, Boulder, Colorado, 1978.

HUTCHINGS, R., Soviet Economic Development, Blackwell, Oxford, 1971.

LAVRISHCHEV, A., Economic Geography of the USSR, Progress Publishers, Moscow, 1969.

MATHIESON, R. S., The Soviet Union: An Economic Geography, Heinemann, London, 1975.

NOVE, A., The Soviet Economy, Allen and Unwin, London, 1961, 2nd edn. 1968.

NOVE, A., An Economic History of the USSR, Lane/Penguin, London, 1969.

SCHWARTZ, H., The Soviet Economy since Stalin, Lippincott, Philadelphia, 1965.

SHABAD, T., Basic Industrial Resources of the USSR, Columbia University Press, New York, 1969.

SHERMAN, H. J., The Soviet Economy, Little, Brown and Co., Boston, 1969.

TREML, V. G., and FARRELL, R. (eds.), The Development of the Soviet Economy: Plan and Performance, Praeger, New York, 1968.

5. Economic sectors

BARR, B. M. and BATER, J. H., "The electricity industry of Central Siberia", Economic Geography, 45, 1969, 349–369.

CAMPBELL, R. W., The Economics of Soviet Oil and Gas, John Hopkins, Baltimore, 1968.

DERBINOVA, M., "Industry located by plan", Geographical Magazine, XLVIII(6), March 1976, 336–341.

DIENES, L., Locational Factors and Locational Developments in the Soviet Chemical Industry, University of Chicago, Dept. of Geography, 1969.

DIENES, L. and SHABAD, T., The Soviet Energy System, Wiley, New York, 1979.

ELLIOT, I. F., The Soviet Energy Balance, Praeger, New York, 1974.

HEBDEN, R. E., "Trends in Soviet trade since 1960", Geography, 65(1), 1960, 49–52.

HEMY, G. W., The Soviet Chemical Industry, Hill, London, 1971.

HODGKINS, J. A., Soviet Power-Energy Resources, Production and Potentials, Prentice-Hall, London, 1961.

LAIRD, R. D., HADJA, J. and LAIRD, B. A., The Future of Agriculture in the Soviet Union and Eastern Europe, Westview Press, Boulder, Colorado, 1977.

LYDOLPH, P. E. and SHABAD, T., "The oil and gas industries of the USSR", Annals Assoc. Amer. Geographers, 50(4), 1960, 461–486.

RATNIEKS, H., "The greatest oil producer", Geographical Magazine, XLVII(10), July 1975, 607–608.

RATNIEKS, H. "USSR switches to eastern gas", Geographical Magazine, XLVIII(7), April 1976, 397–398.

RATNIEKS, H., "Oil shale and the USSR", Geographical Magazine, L(1), October 1977, 46.

RATNIEKS, H., "Soviet hot spots" (subterranean energy), Geographical Magazine, L(11), August 1978, 708–710.

RATNIEKS, H., "Soviet nuclear energy without restraint", Geographical Magazine, LII(1), October 1979, 1–3.

SHABAD, T. and LYDOLPH, P. E., "The Chemical industries in the USSR," Tijdschrift voor Econ. en Soc. Geografie, 1962, 167–179.

SHIMKIN, D. B., Minerals: A Key to Soviet Power, Harvard University Press, Cambridge, Mass., 1953.

STRAUSS, E., Soviet Agriculture in Perspective: A Study of its Successes and Failures, Allen and Unwin, London, 1969.

SYMONS, L. J., Russian Agriculture, Bell, London, 1972.

SYMONS, L. J., "Soviet civil air services", Geography, 58(4), 1973, 328–330.

SYMONS, L. J. and WHITE, C., Russian Transport: An Historical and Geographical Survey, Bell, London, 1975.

6. Regional and local studies

AGRANAT, G. A. "Exploiting and conserving the rich Soviet north", Geographical Magazine, XLVIII(10), 618–622.

ARMSTRONG, T., ROGERS, G. and ROWLEY, G., The Circumpolar North, Methuen, London, 1978.

BOOTH, B. and MORRIS, A. J. W., "Tashkent – City of seven earthquakes", Geographical Magazine, XLVI(8), May 1974, 409–416.

CONOLLY, V., Beyond the Urals – Economic Developments in Soviet Asia, Oxford University Press, London, 1967.

CONOLLY, V., Siberia Today and Tomorrow, Collins, London, 1975.

DEWDNEY, J. C., "The Baykal-Amur Mainline (BAM) project", Geography, 63(2), 1978, 119–122.

DIENES, L. "Basic industries and regional economic development: the Soviet south", Tijdschrift voor Econ. en Soc. Geografie, 1977, 2–15.

GIESE, E., "Transformation of Islamic cities in Soviet Middle Asia into socialist cities", in French and Hamilton, op. cit. 1979.

GORMSEN, E., HARRISS, B. and HEINRITZ, G. "Free enterprise in the USSR" (Moscow's food markets), Geographical Magazine, XLIX(6), March 1977, 376–382.

HAMILTON, F. E. I., The Moscow City Region (Problem Regions of Europe series), Oxford University Press, London, 1976.

KIRBY, E. S., "Where USSR backs USA", Geographical Magazine, XLIV(2), November 1971, 121–123.

KIRBY, E. S., The Soviet Far East, Macmillan, London, 1971.

LAPPO, G., Moscow, Capital of the Soviet Union, Progress Publishers, Moscow, 1976.

MATLEY, I. M., "The Golodnaya steppe: a Russian irrigation venture in Central Asia", Geographical Review, 60(3), 1970, 328–346.

MATRUSOV, N., "Western Siberia offers energy", Geographical Magazine, XLVII(9), June 1976, 548–552.

McBRIDE, B. St. C., "Life in a Moscow suburb", Geographical Magazine, XLI(7), April 1969, 534–539.

NOVE, A. and NEWTH, J. A., The Soviet Middle East: A Communist Model for Development, Allen and Unwin, London, 1967.

POULTON, M. and POULTON, R., "Two Central Asian cultures", Geographical Magazine, XLIX(11), August 1977, 699–704.

RETEYUM, A., "Transformation scene in Central Asia", Geographical Magazine, XLVIII(11), August 1976, 682–686.

ROZANOV, B. G., "Soviet deserts are reclaimed", Geographical Magazine, L(3), December 1977, 162–166.

SALLNOW, J., "Kurile are problem islands", Geographical Magazine, LI(11), August 1979, 729–731.

SHABAD, T. and MOTE, V. L., Gateway to Siberian Resources (The Baykal-Amur Mainline), Wiley, New York, 1977.

SHAW, K., "Situations vacant in Western Siberia", Geographical Magazine LI(2), November 1978, 154–158.

SMITH, D. M., "Siberian city of Science". (Akademgorodok), Geographical Magazine LI(3), 238–242.

THIEL, E., The Soviet Far East, Methuen, London, 1967.

In addition to the works listed above, reference should be made to the journal Soviet Geography; Review and Translation published monthly (except July and August) by Winston, Silver Spring, Maryland U.S.A. in cooperation with the American Geographical Society and to the bibliographies and other works listed in

HARRIS, C. D., Guide to Geographical Bibliographies and Reference Works in Russian or on the Soviet Union, University of Chicago, Dept. of Geography, 1975.

INDEX

Figures in *italics* refer to a map or diagram on the specified page

Abakan *99*
Abaza 57, *99*
Abkhaz ASSR 22, *23*, 100, *101*
Abkhaz people 35
Achinsk 65, *99*
Achisay 49, 65
administrative divisions 22, *23*, *25*, 28
Adygey A.Ob. 22, *23*, 100, *101*
Adygey people 35
Adzhar ASSR 22, *23*, 100, *101*
Afghanistan 2, 26, 52, *105*
Aga-Buryat A.Ob. 22, *23*, *99*
agricultural land 40, *41*, 42
agricultural machinery 60, 62, *63*, 88, 94
agricultural production 46, *47*
agricultural regions *44*, *45*
agriculture 40, *41*, 42, *43*, 44, *45*, 46, *47*, 64, 92, *101*, 104
airways 78, 90
Aktyubinsk 57, 102, *103*
Ak Tyuz *105*
Alapayevsk 57, 96, *97*
Alaska 2, 18, 26, *27*
Alaverdi 49, 58, *59*, 65, 100, *101*
Alay mts. 6
alcohol 42, 64, 66
Aldan r. 8, 80, *81*
Aldan shield 4
alder 16
Aleksandriya *61*
Alekseyevka *61*
Alga 65
alimentary industries *61*, 62, 74, *75*, 90, 92, 94, 100, 104
Alma-Ata 36, *37*, 60, *61*, 71, *103*, 104
Almalyk 49, 58, *59*, 104, *105*
Almetyevsk 95
alpine folds 4
Altay kray *99*
Altay mts 6, *19*
Altay people 34, *35*
Altyn Topkan *105*
aluminium 48, *49*, 58, *59*, 87, 88, *89*, 96, *97*, 100, *101*, 102, *103*, 104, *105*; *see also* alunite, bauxite, kaolinite, nephelite
Alunitdag 100, *101*
alunite 58, 100
America, United States of 2, *3*, 26, 46, 52, 54, 59, 82, *83*, 96
ammonia 66, 104
Amu Darya r. 8, *9*, 104, *105*
Amur oblast *99*
Amur r. 8, *9*, 26, 55, 80, *81*
Amur-Maritime area 18, *19*

Amur-Ussuri lowland 7, 42
Anabar shield 4
Anadyr r. & basin *9*, 51
Anchorage, Alaska 2
Andizhan 28, 30, 71, *105*
Angara r. 8, 48, 54, *55*, 56, *81*, *99*
Angarsk 65, 71, 98, *99*
Angren 104, *105*
antimony 49, 87, 102, *103*
Anzhero-Sudzhensk *61*, *99*
apatite 65, 66, 87, *89*
Apatity 88, *89*
arable land 46, *47*
Aramil 57, 67, *69*, *97*
Araks r. 8
Aral Sea 6, 8, *9*, 74, *103*, *105*
Aralsulfat *103*
Arctic Ocean 2, 6, 7, 8, *9*, 10, 12, 52, 74, 80
Arkalyk *49*, 102
Arkhangelsk *37*, *61*, 62, *63*, 71, 72, 89
Armavir 71
Armenia, Armenian SSR 22, *23*, 24, 32, 74, 100, *101*
Armenian highlands *17*
Armenian people 34, *35*
Artem *99*
Artemovsk 48, *49*, 92, *93*
Artemovskiy *61*, *97*
Arzamas *69*, 71
Asbest *49*, *97*
asbestos 48, *49*, 87, *97*, 103
ash 16
Asha 57, *97*
Ashkhabad *37*, *69*, 71, *105*
aspen 16, *17*
Astrakhan 10, 24, *37*, 62, *63*, 71, 94, *95*
Atasu *49*, 57, 102
Atbashi *105*
Atlantic Ocean 34, 74
Australia 2
automobiles *see* motor vehicles
Autonomous Oblasts (A.Ob's) 22, *23*
Autonomous Okrugs (A.Ok's) 22, *23*
Autonomous Soviet Socialist Republics (ASSR's) 22, *23*
Ayaguz *69*
Azerbaydzhan, Azerbaydzhanian SSR 22, *23*, 24, 32, 100, *101*
Azerbaydzhanian people 34, *35*
Azov, Sea of 4, 8, 56, 92, *93*, 100

Bakal 57, 96, *97*
Baku *13*, 36, *37*, 52, 53, 60, *61*, 63, 65, 67, *69*, 71, 100, *101*
Balakovo *61*, 65, 67, 94, *95*
Balashikha *69*
Baley *49*
Balkans 48
Balkhash *49*, 74, 102, *103*
Balkhash, lake *103*
Baltic languages, peoples 34, *35*

Baltic economic region, republics 12, *23*, 24, *25*, 26, *29*, 32, 34, 46, 56, 72, *73*, 78, *87*, 88, *89*; *see also* Estonia, Latvia, Lithuania
Baltic Sea 2, 8, *9*, 62, 80
Baltic shield 4, 48
Bangladesh *3*
Barents Sea *9*, *89*
barley 40, *41*, 44
Barnaul *37*, *61*, 63, 67, *69*, 71, *99*
Barysh *69*
Bashkir ASSR 22, *23*, 48, 66, 94, *95*, 96
Bashkir people 34, *35*
Baskunchak, lake 94, *95*
Batamshinskiy *103*
Batumi *13*, 65, 100, *101*
bauxite 48, *49*, 58, 88, 96, 102
Baykal, lake 6, 8, 54, 74, 76, *99*
Baykal-Amur Mainline (BAM) railway 58, 76, *77*, 98, *99*
Baykalia 18, *19*
Baymak *49*
Bayram Ali *69*, *105*
Bazay *103*
beech 16
beef 44
Bekabad 56, 57, 104, *105*
Bekdash *105*
Belaya r. 48, 94
Belaya Kholunitsa *61*
Belaya Tserkov 67, *93*
Belgium *83*
Belgorod *61*, *91*
Belogorsk *49*, 63
Belorussia, Belorussian economic region, SSR 8, 12, 22, *23*, 24, *25*, *29*, 40, 42, 53, 60, 62, 66, 67, 68, 72, 86, *89*
Belorussian people 34, *35*, 88, 102
Belorechensk 65
Beloretsk 57, *95*
Belovo 58, *59*, *99*
Belozerka 56, 57, *93*
Beltsy 71
Bendery *69*
Berdichev *69*
Bereza 88, *89*
Berezenskaya gasfield *101*
Berezniki 48, *49*, *59*, 65, 66, 71, 96, *97*
Berezovo *99*
Bering Sea *9*
Bering Strait 14, 26
Bezhetsk 63
Bilimbay 57, *97*
birch 16, *17*
Birobidzhan 63, 71, *99*
birth rate 32, *33*
Biysk *61*, *69*, 71, *99*
black earth 14, *15*
Black Earth Centre economic region 24, *25*, *29*, 40, 46, 48, *73*, 74, 86, *87*, 90, *91*

Black Sea 2, 4, 8, 9, 10, 12, 26, 28, 52, 62, 80, *93*, 100
Blagoveshchensk *61, 63, 71, 99*
Bobruisk *67*
Bokhara *see* Bukhara
Boksitogorsk *49*, 58, 88, *89*
Boksol *49*
Borisov *67*
Borovichi *69, 71*
boundaries 2, 22, *23, 24, 25,* 26, *27,* 87
Bozsu r. 104, *105*
Bratsk 54, *59*, 98, *99*
Brazil *3*
Brest *69, 89*
brown forest soils 14, *15*
Bryansk 57, *63, 65, 69, 91*
buckwheat 40
Bug, Southern, r. 8, *93*
Buguruslan *97*
building materials 76, *77,* 80, *81*
Bukachacha coalfield *51,* 98, *99*
Bukhara *69, 71,* 104, *105*
Bukhtarma *103*
Bulgaria 82, *83*
Bulgars 34, *35*
Bureya r. & basin *51, 57*
Buron *101*
Buryat ASSR 22, *23,* 98, *99*
Buryat people 34, *35*
buses *78,* 88
butter 74
Buzuluk *61, 71*

Caledonian folds 4
Canada 2, *3,* 72
canals 80, *81,* 90, *105*
canning 74, *75*
capital equipment 68
Carpathian mts 6, *19*
Caspian lowland 6, *19,* 102
Caspian Sea 2, 4, 6, 8, 52, 54, 62, 74, 80, 100, *101,* 104, *105*
cattle 44
Caucasia, Caucasus mts, region 4, 6, 8, 16, *19,* 22, 26, 44, 45, 54, 76, *101; see also* North Caucasus, Transcaucasia
Caucasian languages, peoples 34, *35*
cellulose 72, 104
cement 104
Central Asia, Central Asian economic region 8, 10, 12, 16, 24, *25,* 26, 28, *29,* 30, 32, 34, 36, 40, 42, 44, 45, 46, *47,* 48, *51, 52, 53,* 54, 58, 60, 62, 64, 68, 74, 76, 80, 87, 90, 92, 94, 96, 102, 104, *105; see also* Kirgizia, Tadzhikistan, Turkmenia, Uzbekistan
Central Asian lowland 4, 104
Central Russian elevation 6, 18
Central Siberian plateau 6, 8, 18, *19*
Centre, Central economic region 24, *25, 27,* 28, 30, 32, 36, 40, 42, 46, 54, 60, 62, 64, 66, 68, *73,* 74, 86, *87,* 90, *91,* 94, 104

cereals 40, *41, 42,* 44, 46, *47, 101; see also* grain
Changytarsh *65*
Chapayevsk *65*
charcoal 56
Chardara *103, 105*
Chardzhou *65, 69,* 104, *105*
Charvak *105*
Chauvay *105*
Cheboksary 37, *63, 69, 91*
Chechen-Ingush ASSR 22, *23,* 100, *101*
Chechen people *35*
Cheleken *105*
Chelyabinsk 36, *37,* 57, 62, *63, 69, 71,* 96, *97*
Chelyuskin, Cape 2
chemical industries 52, *61,* 62, 64, *65,* 66, *67,* 72, 88, 90, 92, 94, *96,* 104
Cheremkhovo *61,* 98, *99*
Cherepovets *37,* 56, *57, 65,* 88, *89*
Cherkassy 66, *67, 93*
Cherkess people *35*
Chernigov *67, 69, 93*
Chernovtsy *61, 69, 71, 93*
chernozem soils 14, *15,* 90, 92
chestnut soils 14, *15*
Chiatura 48, *49, 57,* 100, *101*
China 2, *3,* 26, 76, 80, 104, *105*
Chimkent *37,* 58, *59, 65, 69, 103, 105*
Chirchik *63, 65,* 104, *105*
Chirkey r. 100, *101*
Chirkoy *71*
Chita *37, 59, 61,* 98, *99*
chlorine 66
Choruk Dayron *105*
christianity 34
chrome 48, *49,* 54, *87, 97,* 102, *103*
Chu r. 8, *103, 105*
Chukchi people *35*
Chukot A.Ok. 22, *23, 99*
Chukot basin *51*
Chukot peninsula 2
Chusovoy *57, 97*
Chuvash ASSR 22, *23, 91*
Chuvash people *35*
citrus fruits 42, *43,* 44, *45,* 100
Civil War 32
climate 10, *11,* 12, *13*
clothing 64, 66, 70, *71*
coal *49,* 50, *51, 53,* 54, 56, 64, 66, 76, *77,* 80, *81, 87,* 89, 90, *91,* 92, *93,* 96, *97,* 98, *99,* 100, *101, 102, 103,* 104, *105; see also* lignite
cobalt 48, *49, 87,* 96, *97*
combine harvesters 60, 62, *63*
COMECON (Council for Mutual Economic Assistance; CMEA) 2, *3,* 52, 54, 56, 82
Communism, mt. 6
coniferous forest 16, 72, *see also tayga*
consumer goods 60, 68, 70

copper 48, *49, 59, 87, 89,* 94, *95, 96, 97, 99,* 100, *101, 102, 103,* 104, *105*
cotton 40, 42, *43,* 44, *45,* 68, *69,* 100, 104
Crimea 10, *19, 53,* 92, *93*
crops 40, *41, 42, 43*
Cuba 82, *83*
Czechoslovakia 2, 26, 82, *83*

dacron 66
Dagestan ASSR 22, *23,* 100, *101*
Dagestan, peoples of *35*
Dahuria *19*
dairying, dairy products 44, 74, *75*
Dankov *67*
Danube r. *93*
Darasun *61*
Darvaza *105*
Dashkesan *49, 57, 101*
Daugavpils *67, 71*
death rate 32, *33*
deciduous forest 16, *17, 19,* 92, *95*
Dedovsk *69*
Degtyarsk *97*
Deputatskiy *49*
Derbent *69*
desert 6, 12, 14, *15,* 16, *17, 19,* 30, *95*
diamonds 48, *49*
diet 46, 74
Diomede Is. 2
Dmitrov *71*
Dnepr r. 6, 8, 54, *55,* 56, *57,* 64, 76, 80, *81,* 88, *89, 93*
Dnepr-Bug canal 80, *81, 89*
Dneprodzerzhinsk *57,* 92, *93*
Dnepropetrovsk 36, *37,* 60, *61, 67, 71,* 92, *93*
Dnestr r. 8, *93*
Dolgano-Nenets (Taymyr) A.Ok. 22, *23*
Dolgany people *35*
Don r. 6, 8, *9, 55, 81, 91, 95,* 100, *101*
Donbass (Donets Basin) 24, 36, 48, 50, *51,* 54, 56, *57,* 58, 60, 64, 76, *87,* 92, *93,* 98, *101*
Donets-Dnepr economic region 24, *25, 29,* 30, 48, 86, 92, *93*
Donets Heights 6
Donetsk 28, *37, 57, 61, 69, 71,* 92, *93*
Donets r. *93, 101*
Dorogobuzh *65,* 66
drainage 8, *9,* 14, *18,* 44
Dubenskiy *97*
Duchy of Warsaw 26, *27*
Dunavetsy *69*
Dushanbe *37, 69, 105*
Duvannyy *101*
Dvina, Northern, r. 8, *9,* 54, *55,* 80, *81*
Dvina, Western, r. 8, *55, 81,* 88, *89*
Dzerzhinsk *65, 67*
Dzhambul *49, 59, 69,* 102, *103*
Dzharkak *105*
Dzhezdy *49, 57*

Dzhezkazgan *49, 59,* 102, *103*
Dzhetygara *103*
Dzhizak oblast *105*
Dzungarian Gate 6

East European plain, platform 4, 6, 12, 16, 18, *19,* 28
East Kazakh oblast *103*
East Prussia 26, *27*
East Siberian economic region 10, 12, 24, *25,* 29, 30, 32, 40, 46, 48, 56, 58, 72, 76, *87,* 98, *99*
East Siberian Sea *9, 99*
Echmiadzin 67
economic regions 24, *25,* 86
Ege-Khaya *49*
Ekibastuz coalfield 50, *51,* 54, 102, *103*
Elbruz, Mt. 6
electrical equipment 60, *61,* 88
electricity 50, 52, 55, *87,* 89, 90, *91,* 93, *95,* 97, 98, *99,* 101, *103,* 104, *105; see also* hydro-electricity
Elektrostal 57, 60, *61, 91*
Elista *95*
elm 16
Elton, lake 94, *95*
Emba (Guryev) oilfield *53,* 96, 102, *103*
Emba r. *103*
Engels *63, 67, 71*
engineering 60, *61,* 62, *63,* 67, *71,* 88, 92, 94, 96, 98, 104
Estonia, Estonian SSR 22, *23,* 24, 66, 88, *89*
Estonian people 34, *35*
ethnic groups 34, *35*
European Economic Community *3,* 74
Eveni people *35*
Evenki A.Ok. 22, *23,* 28
Evenki people 22
evaporation 14

Fabrichniy *69*
fallow 46
Far Eastern economic region 6, 7, 8, 10, 12, 14, 16, 24, *25,* 29, 30, 32, 40, 42, 44, 46, 48, 56, 58, 70, 72, 86, *87,* 98, *99*
Farkhad *105*
Fennoscandian shield 4
Fergana, Fergana basin *45, 53, 65,* 66, *67, 69,* 104, *105*
ferro-alloys 48, 57, 86, *87, 93,* 96, *97, 101,* 102, *103*
fertilisers 44, 64, *65,* 66, 76, *77,* 88, 104
Fiat Motor Co. 62
Finland *2,* 4, 26, *27,* 82, *83,* 88
Finland, Gulf of *89*
Finno-Ugrian languages, peoples 34, *35*
fir 16
firewood 72
fish 74, *75,* 94
flax 40, 42, *43,* 44, 68, 88
flour 74, *75*
flourspar *49*
fodder crops 40, 42, 44, 46, 47
food processing *see* alimentary industries

footwear 70, *71*
forests, forestry 16, *17,* 18, *19,* 40, 46, 47, 72, *73; see also* timber
Fosforitny *65*
France 82, *83*
frost 12, *13*
fruit 44, 45, 74, *75,* 100
Frunze 37, *63,* 68, *69, 71,* 104, *105*
Fryanovo *69*
Furmanov *71*

gas, natural 33, *49,* 50, 52, *53,* 54, 64, 66, 82, *87,* 88, 90, 92, *93, 94, 95, 96, 97,* 98, *99,* 101, 102, *103,* 104, *105*
Gaurdak *49,* 64, *65,* 104, *105*
Gavrilov Yam *69*
Gay *49, 97*
Gaz Achak *105*
Gazli *105*
generators 60
geology *4, 5*
Georgia, Georgian SSR 22, *23,* 24, 32, 48, *51,* 62, 100, *101*
Georgian people 34, *35*
Germans 34, *35*
Germany 54, 82, *83,* 90
Glubokoye *59, 103*
gold 48, *49,* 96, *97,* 98, *99,* 102, *103*
Gomel 37, *65, 71, 89*
Gori *69*
Gorkiy 28, 36, *37,* 57, *59,* 62, *63, 65, 69, 71,* 90, *91*
Gorlovka 37, *61, 65, 93*
Gorno-Altay A.Ob. 22, *23, 99*
Gorno-Badakhshan A.Ob. 22, *23,* 30, *105*
Gorodets *91*
Gorokhovets *61*
grain 47, 64, 74, 76, *77,* 80, *81,* 82, 102, 104
graphite 48, *49*
grazing 44, 46
Greeks *35*
Greenland 2
grey forest soils 14, *15*
Grodno *65,* 66, *67, 71, 89*
growing season 44
Groznyy 37, 52, *65, 71,* 100, *101*
Gubakha *97*
Gubkhin 57, *99*
Gugurtli *105*
Guryev 36, *65, 67, 103*
Guryev (Emba) oilfield 52, *53,* 96, 102, *103*
Guryevsk *99*

hardwood 72, *73*
hemp 40, *43,* 44, *45,* 68, *69*
holly 16
hornbeam 16
Hudson's Bay 2
humus 14
Hungary *2,* 6, 58, 82, *83*
hunting 44

Hercynian folds 4
hydro-electricity *53,* 54, *55,* 56, 58, *59,* 72, 80, 86, *87,* 88, *89,* 90, *91,* 92, *93,* 94, *95,* 96, *97,* 98, *99,* 100, *101,* 102, *103,* 104, *105*

Igrim *99*
Ili r. *103, 105*
India *2, 3,* 26
Indigirka r. 8, *9*
Indo-European languages, peoples 34, *35*
Indo-Iranian languages, peoples 34, *35*
Indonesia *3,* 49
industrial crops 40, 42, 47
Ingichka *49, 105*
Inguri r. 100, *101*
Ingush people *35*
inland drainage 8, *9*
intrusive rocks *4, 5*
Ionava 64, *65*
Iran *2,* 6, 26, 52
Iraq *83*
Iriklinskiy *97*
Irkutsk 37, *61, 67, 71,* 72, 98, *99*
Irkutsk basin *51*
iron 48, *49,* 56, 57, *87,* 88, *89,* 90, *91,* 92, *93, 94, 95,* 96, *97,* 98, *99,* 100, *101,* 102, *103,* 104
irrigation 40, 42, 44, 54, 92, *93, 95,* 100, *101,* 104, *105*
Irtysh r. & basin 54, *55,* 80, *81,* 102, *103*
Irtysh–Karaganda canal 102, *103*
Ishim r. *103*
Ishimbay *65*
Issyk-Kul, lake *105*
Issyk-Kul, oblast *105*
Italy 82, *83*
Ivano-Frankovsk *71,* 92, *93*
Ivanovo 37, *61,* 68, *69, 91*
Ivantayevka *69*
Izhevsk 57, *63, 97*

Japan *2, 3,* 26, 74, 82, *83,* 96
Japan, Sea of *9*
Jewish (Yevreysk) A.Ob. 22, *23, 99*
Jews 34, *35*
jute 68, *69*

Kabardino-Balkar A.Ob. 22, *23,* 100, *101*
Kabardin people *35*
Kachkanar 48, *49,* 56, *57,* 96, *97*
Kachug *63*
Kadiyevka 57, 60, *61, 93*
Kadzharan *49,* 100, *101*
Kafan 100, *101*
Kalinin *63, 67, 69, 71, 91*
Kaliningrad *2,* 37, 62, *63,* 88, *89*
Kalmas *101*
Kalmyk ASSR 22, *23, 95*
Kalmyk people 34, *35*

Kaluga *37, 61, 63, 91*
Kalush *65*
Kalyazin *71*
Kambarka *63*
Kama r. 8, 18, 54, 55, 80, *81*, 94, 95
Kamchatka 7, *19*, 30, 51, *98*, *99*
Kamensk-Shakhtinskiy *61*
Kamensk-Uralskiy *61, 67*
Kamyshidza *105*
Kamyshin *71, 73*
Kamyshlov *71*
Kandalaksha *58, 59, 89*
Kansk *69, 99*
Kanev *93*
Kansk-Achinsk coalfield *50, 51*, 54, *99*
Kaolinite *58*, 104
Kapchagay *103*
kapron 66, 67
Kara Balty *105*
Karabash *59, 96, 97*
Kara Bogaz Gol *105*
Karachayevo-Cherkess A.Ob. 22, 23, *100, 101*
Karachayev people 35
Karadag *101*
Karaganda *37*, 50, 51, 56, 57, *61, 62, 67, 71*, 76, *102, 103*, 104
Kara-Kalpak ASSR 22, 23
Kara-Kalpak people 35
Kara Kum canal 80, *81*
Kara Sea 9, *89, 99*
Karatau *49*, 64, *65*, *102, 103*, 104
Karazhal *49, 57*, 102
Karelia, Karelian ASSR 4, 22, *23*, 54, 58, 74, 88, *89*
Karelian people 35
Karlyuk *65*
Karpinsk *97*
Karsakpay *59*
Kashkadarya oblast *105*
Kattakurgan *69*
Kaunas *37, 61, 67, 69*
Kazakh people 34, *35*
Kazakhstan, Kazakh SSR, economic region 14, 16, 22, *23*, 24, *25*, 28, *29*, 30, 32, 36, 40, 42, 46, *47*, 48, *50*, 54, 56, 58, *59*, 60, 64, *72*, 76, 86, *87*, 94, 96, *102, 103*, 104, *105*
Kazakh upland 4, 6, *19*, 102
Kazan 36, *37, 67, 91*, 94, 95
Kedainyai *65*
Kem r. 88, *89*
Kemerovo 30, *37, 67, 99*
Kentau *103*
Kerch *49, 57*, 92, *93*
Khabarovsk *37, 61, 63, 65, 71, 98, 99*
Khakass people 34, *35*
Khamza *65*
Khanty-Mansiy A.Ok. 22, *23*, 30
Khanty people 34, *35*
Khapcheranga *49*

Kharkov 36, *37, 61, 62, 63, 67, 69*, 92, *93*
Khartsyzsk *57, 93*
Khaydarkan *105*
Kherson *37, 62, 63, 65, 93*
Khibiny mts. 6
Khimkan *49, 57*
Khimki *91*
Khmelnitskiy *93*
Khodino *88, 91*
Khorezm *105*
Khorog *105*
Khromtau *49, 103*
Khrushchev, N.S. 40
Kiev 26, 28, 32, 36, *37, 61, 63, 67, 71*, 92, *93*
Kievan Rus 26
Kimry *69*
Kineshma *69*
Kingisepp *65*
Kirpichli *105*
Kirgizia, Kirgiz SSR 22, *23*, 24, 68, *104, 105*
Kirgiz people 34, *35*
Kirishi 64, *65, 89*
Kirov *37, 61, 65, 67, 71, 72*, 90, *91*
Kirovabad *65, 69*
Kirovakan *65, 67*
Kirovgrad *49, 59, 96, 97*
Kirovograd *71, 93*
Kirovsk *49*, 66, *88, 89*
Kirzhach *69*
Kiselevsk *61, 99*
Kishinev *37, 63, 71, 93*
Kizel *51, 57, 97*
Kizilyar *65*
Klaypeda *63, 69, 89*
Klin *67*
Klintsy *69*
Klyuchevskiy, Mt. 7
knitwear *71*
Kohtla Jarve 64, *65*
Kokand *65, 69, 105*
Kokchetav *103*
Kola peninsula 2, 6, 14, 48, 66, *89*
Kolchugino *71*
Kolomna *63, 91*
Kolpino *61*, 77, *89*
Kolyma r. & basin 7, 8, 9, 80, *81*
Komi ASSR 22, *23*, 88, *89*
Komi people 34, *35*
Komi-Permyak A.Ok. 22, *23, 97*
Komi-Permyak people 34, *35*
Komi-Ukhta oilfield *52, 53*, 88, *89*
Kommunarsk *57, 93*
Komsomolsk-na-Amure *37, 56, 57, 61, 65, 71, 98, 99*
Konstantinovka *57, 58, 59, 65, 93*
Kopeysk *61, 97*
Korea 2
Koryak A.Ok. 22, *23, 99*

Koryak people 34, *35*
Kosaya Gora *57*
Kostroma *61*, 68, *69, 91*
Kotlas *63*, 88, *89*
Kounradskiy *102, 103*
Kovda r. 88, *89*
Kovdor *49*, 88
Kovrov *69*
Koytash *105*
Kramatorsk *57*, 60, *61, 93*
Krasnoarmeysk *69*
Krasnodar *37*, 52, *65, 69, 71*, 100, *101*
Krasnogvardenskiy *61*
Krasnooktyabrskiy *49, 102, 103*
Krasnoturinsk *58, 59, 96, 97*
Krasnouralsk *49, 59, 65, 96, 97*
Krasnovodsk *65, 105*
Krasnoyarsk *37*, 54, 56, *59, 61, 63, 67*, 68, *69, 71, 72, 98, 99*
Krasnyy Luch *61, 93*
Krasnyy Sulin *57, 93, 101*
kray 25
Kremenchug 56, *57, 65, 71, 93*
Krivoy Rog *37*, 48, *49*, 56, *57, 65, 93*
Kuban r. & plain 8, 42, 74, 100, *101*
Kugitang *105*
Kulebaki *57, 91*
Kulyab oblast *105*
Kuma r. 88, *89*
Kumak *97*
Kumdag *105*
Kungur *71, 97*
Kura r. 100, *101*
Kurgan *37, 63, 71, 97*
Kurgan-Tyube oblast *105*
Kurovskoye *69*
Kursk *37, 67, 71, 91*
Kursk Magnetic Anomaly (KMA) 48, *49*, 56, *57, 91*
Kusa *61*
Kushva *57, 97*
Kustanay 48, *67, 71*, 96, *102, 103*
Kutaisi *63, 69*
Kuybyshev 32, 36, *37*, 54, *63, 65, 67, 69, 73*, 94
Kuzbass (Kuznets Basin) 30, 36, *49*, 50, 54, 56, *57, 58, 61*, 64, 66, 76, 96, *98, 99*, 104
Kuznetsk *71*
Kyshtym *59, 61*, 96, *97*
Kyzyl *71, 99*
Kyzyl-Dzhar *105*
Kzyl-Orda *103*
Kzyl Tumshuk *105*

labour 60, *99*
Ladoga, lake *89*
land use 40, 42, 46, 47
languages 34
Laptev Sea 9, *99*
larch 16

lathes 60
Latvia, Latvian SSR 22, 23, 24, 62, 89
Latvian people 34, 35
lavsan 66, 67
lead 48, 49, 58, 59, 87, 96, 97, 98, 99, 100, 101, 103, 104, 105
leatherwork 71
legumes 40, 42
Lena r. & basin 4, 6, 8, 9, 12, 16, 44, 50, 51, 52, 53, 55, 80, 81, 99
Lenin. V. I. 54
Leninabad 69, 71, 105
Leninakan 69, 71
Leningrad 2, 12, 13, 24, 28, 32, 36, 37, 57, 58, 60, 61, 63, 65, 67, 69, 71, 76, 88, 89
Leningradskaya gasfield 101
Leninogorsk 58, 59
Lenin Peak 6
Lesnogorsk 67
lianas 16
lichens 18
light industry 60, 68, 70, 88
lignite 49, 50, 51, 53, 54, 64, 87, 90, 91, 93, 96, 98, 99, 101, 102, 103
Likino-Dudyevo 63
lime 44
linen 68, 69
Lipetsk 37, 56, 57, 63, 65, 90, 91
Lipovtsy 99
Lisakovsk 57, 103
Lisichansk 64, 65, 93
Lithuania, Lithuanian SSR 22, 23, 24, 32, 66, 89
Lithuanian people 34, 35
livestock 40, 46, 47, 100, 101
Liyepaya 56, 57, 88, 89
Lubna 91
lucerne 44, 45
Luchegorsk 99
Lutsk 93
Lvov 37, 61, 63, 71, 92, 93
Lovo-Volyn coalfield 51
Lysva 57, 97
Lyubertsy 63, 67, 69, 91
Lyudinovo 63

Maardu 65
machinery 59, 60, 61, 62, 83
machine tools 60, 61, 62, 88, 96
Madneuli 49, 100, 101
Magadan 51, 71, 88, 89
magnesite 48, 49
Magnitorgorsk 37, 48, 49, 50, 56, 57, 61, 67, 71, 76, 96, 97, 102
maize 40, 41, 42, 44, 100
Makeyevka 37, 57, 93
Makhachkala 69, 101
Malaysia 48
Manchuria 2
manganese 48, 49, 56, 57, 87, 92, 93, 96, 97, 100, 101

Mangyshlak oblast, oilfield 53, 96, 102, *103*
Mansiy people 34, 35
maple 16
Margilan 69, 105
Maritime (Primorskiy) kray 30, 98, 99
Mariy ASSR 22, 23, 24, 91
Mariy oblast 69, 105
Mariy people 34, 35
Marshansk 69
Maykop 52, 100, 101
Maykuben 103
Mayskoye 105
Mazeykiay 65
meadows 46, 47
meat 47, 74, 75
Mediterranean Sea 80
Mednogorsk 49, 59, 97
Melekess 69, 71
Melenki 69
Meleuz 65
Melitopol 63, 93
mercury 48, 49, 87, 105
Mesozoic rocks 4
Messoyakhan 99
metallurgical equipment 60, 61
metallurgy 58, 59, 64, 86, 87, 89, 91, 93, 94, 95, 96, 97, 98, 99, 101, 102, 103, 105
Mezhdurechensk 99
Miass 57, 62, 63, 97
mica 48, 49
Middle East 52
Middle Siberian railway 76
migration 32, 90, 104
milk 47, 74
millet 40, 44
Minchegaur 69, 101
minerals 48, 57, 64, 66, 76, 77, 80, 81, 86, 87, 89, 91, 93, 94, 95, 96, 97, 99, 101, 103, 104, 105
mineral salts *see* salts
mining equipment 60, 61
Min Kush 105
Minsk 32, 36, 37, 61, 63, 69, 71, 88, 89
Minusinsk basin 51, 98
Mirniy 49
mixed forest 16, 17, 19, 28, 44, 94, 95
Mogilev 37, 63, 67, 71, 89
Moldavia, Moldavian SSR 8, 16, 22, 23, 24, 25, 28, 29, 30, 32, 40, 42, 44, 46, 62, 74, 86, 87, 92
Moldavian people 34, 35
molybdenum 48, 49, 87, 98, 99, 100, 101, 103, 104, 105
Monchegorsk 49, 59, 88, 89
Mongolia 2
Mongolian peoples 34, 35
monsoon 16
Mordov ASSR 22, 23, 68, 91
Mordov people 34, 35
Moscow 2, 12, 13, 18, 26, 28, 30, 32, 36, 37, 56, 57, 60, 61, 62,

Moscow (*contd.*)
63, 64, 65, 67, 68, 69, 71, 76, 78, 86, 90, 91
Moscow basin 4, 50, 51
Moscow canal 80, 81, 91
Moscow r. 80, 91
moslems 32, 34
mosses 16
motor vehicles 62, 63, 88, 94, 96
Mozyr 65, 67, 88, 89
Murmansk 30, 37, 71, 88, 89
Murom 63, 69
Muscovy 26, 27
Myldzhino 99
Mytishchi 63, 67, 91

Naberezhniye Chelny 62, 63, 94, 95
Nadvoitsy 58, 59, 89
Nagorno-Karabakh A.Ob. 22, 23, 100, 101
Nakhichevan ASSR 22, 23, 100, 101
Nakhodka 99
Nalchik 71
Namangan 30, 67, 69, 105
Napoleonic wars 26
Narofominsk 69
Narva 69
Narva r. 8, 88, 89
Naryn oblast 30, 105
Naryn r. 104, 105
nationalities 34, 35
natural gas *see* gas
natural increase 32, 33, 90
natural regions 14, 18, 19
natural vegetation 16, 17, 18, 19
Navoi 65, 66, 67, 104, 105
Nazarovo 99
Nebit Dag 105
Neftekamsk 95
Neftezavodsk 65
Nenets A.Ok. 22, 23, 89
Nentsy people 34, 35
nephelite 88, 100
Nerchinsk, Treaty of 26
Nerekhta 69
Nevinnomyssk 65, 66, 67, 91, 101
newsprint 73
nickel 48, 59, 87, 88, 89, 97, 98, 99, 103
Nigeria 3
Nikel 49, 89
Nikitovka 49
Nikolayev 37, 58, 62, 63, 71, 93
Nikologory 69
Nikopol 48, 49, 56, 57, 92, 93
nitrogen 64, 66, 104
nitron 66, 67
Nizhnekamsk 67, 94, 95
Nizhnevartovsk 99

Nizhniy Sergi 57
Nizhniy Tagil 37, 48, 49, 56, 57, 63, 67, 96, 97
Nizhnyaya Salda 57, 97
Nizhnyaya Tunguska r. 8
Nogantsy people 35
Noginsk 69, 91
nomadism 44
Norilsk 36, 48, 49, 58, 59, 98, 99
North Caucasus economic region 12, 24, 25, 28, 29, 30, 36, 40, 42, 46, 52, 53, 68, 72, 74, 86, 87, 100, 101
North European plain 4
North Kazakh oblast 103
North Osetin ASSR 22, 23, 100, 101
North-western economic region 24, 25, 28, 29, 30, 40, 42, 46, 56, 68, 73, 87, 88, 89
Norway 2, 4
Novaya Zemlya 18
Novgorod 65, 66, 88, 89
Novo-Altaysk 63
Novoblagoveshchensk 65
Novoburyevsk 61
Novocherkassk 61, 62, 63, 67
Novokuybyshevsk 65
Novokuznetsk 37, 57, 59, 99
Novomoskovsk 57, 65, 66, 93
Novopiskovo 69
Novopolotsk 88, 89
Novorossiysk 101
Novoshakhtinsk 93
Novoshakhtinskiy 99
Novosibirsk 36, 37, 57, 59, 61, 69, 71, 98, 99
Novotroitsk 49, 57, 96, 97
Novovyazniki 69
nuclear power 54, 55, 87, 102, 103
Nukus 105
Nura r. 103
Nurek 105
Nyaumunas r. 8
nylon 66
Nytva 57

oak 16
oats 40, 41, 44
Ob r. 8, 9, 54, 55, 80, 81
Obukhovo 69
oblast 24, 25
Odessa 12, 13, 36, 37, 57, 63, 65, 66, 67, 68, 69, 71, 92, 93
Ogodzha 99
oil 33, 49, 50, 52, 53, 54, 64, 66, 76, 77, 80, 81, 82, 87, 89, 90, 92, 93, 94, 95, 97, 98, 99, 100, 101, 102, 103, 104, 105
oil-drilling equipment 60
oil refining 52, 53, 64, 65, 88, 96, 97, 103, 104
oilseeds 40, 100
oilshale 53, 87, 88, 89, 94
Oka r. 68, 80, 81, 90, 91
Okarem 105
Okhotsk, Sea of 9, 52, 98, 99

Oktyabrsk 103
Olavyannaya 61
Olekma r. 8
Olenogorsk 49, 88, 89
Omsk 37, 63, 65, 67, 69, 71, 99
Omutninsk 57, 91
Onega, lake 80, 89
Onega, r. 58, 88, 89
opencast mining 102
Ordzhonikidze 37, 49, 58, 59, 71, 100, 101
Orekhovo-Zuyevo 63, 67, 69, 91
Orel 37, 57, 71, 91
Orenburg 37, 52, 54, 69, 71, 97
orlon 66
Orsha 69, 89
Orsk 49, 59, 60, 61, 65, 71, 96, 97
Osetin people 35
Osh 69, 105
Ossiniki 99
Ostashkov 71
Ozery 69

Pacific Ocean 2, 4, 8, 9, 10, 12, 26, 74, 76
Pakistan 3
Paleo-Asiatic languages, peoples 34, 35
Paleozoic rocks 4, 5
Pamir mts 6
pasture 46, 47
Pavlodar 63, 65, 71, 102, 103
Pavlovo 63
Pavlovskiy Posad 69
Paz r. 89
peat 16, 53, 54, 87, 88, 89, 90, 91
Pechora r. & basin 8, 9, 50, 51, 54, 55, 81, 88, 89
Penza 61, 71, 95
peoples 34, 35
Pereslavl-Zapesskiy 69
Perm 36, 37, 61, 65, 66, 71, 72, 96, 97
permafrost 14, 16
Pervouralsk 57, 65, 97
Peter the Great 26
Petropavlovsk 71, 98, 103
Petrovsk-Zabaykalskiy 56, 57, 99
Petrozavodsk 63, 71, 89
Petsamo 88
phosphates 48, 49, 64, 65, 102, 104
phosphorite 65
Pikalevo 65
pine 16
pipelines 52, 59, 78, 87, 88, 90, 91, 92, 93, 95, 97, 99, 100, 101, 103, 104, 105
Plast 97
plastics 66, 67, 104
platinum 48, 49, 97, 99
Plavinas 89
Plesetsk 49, 88, 89

Plunge 69
Podkamennaya Tunguska r. 8
Podolsk 91
podsol 14, 15, 16
Poland 2, 26, 80, 82, 83, 92
Poles 34, 35
Polesye 18
Polevskoy 57, 61, 97
political divisions 22, 23
Polotsk 64, 65
Poltava 37, 61, 69, 92, 93
Polunochnoye 49, 57, 96, 97
polyvinyl chloride (PVC) 66
population 2, 3, 23, 28, 29, 30, 31, 32, 33, 35, 36, 86, 88, 90, 92, 94, 96, 98, 100, 102, 104
potash 48, 49, 64, 65
potatoes 40, 42, 43, 44, 46, 47, 64
pre-Cambrian rocks 4
precipitation 10, 12, 13, 44, 45
pressure, atmospheric 10, 11
Prikumsk 101
Priluki 67
Primorskiy (Maritime) kray 30, 98, 99
Pripet marshes 80
Pripyat (Pripet) r. 89, 93
Prokopyevsk 37, 71, 99
protein 74
Prussia, East 26, 27
Prut r. 93
Pskov 71, 88, 89
pulp (wood) 72, 73
Punga, r. 99
Pushkino 69
Putoran mts 6
Pyarnu 69
Pyatigorsk 71
pyrites 89

Quaternary rocks 4, 5

railways 72, 76, 77, 78, 87, 88, 89, 91, 93, 95, 97, 99, 101, 103, 105
railway rolling stock 60, 62, 63, 88, 96
rain 12, 13
Ramenskoye 69
Rasskozovo 69, 71
Ratmanov I. 2
rayon 66, 67
Razdan r. 100, 101
Rechitsa 88, 89
Regar 104, 105
regions 18, 22, 24, 86, 87
reindeer 44
relief 6, 7, 14, 18, 91, 95, 97, 101, 103, 105
religion 34
Rettikhovo 99
Revda 97
Rezh 59, 96, 97

rice 40, *41*, 42, 44, 45, 100
Riga *37*, *61*, 62, *63*, 65, 66, 67, 69, 88, *89*
Rioni r. 100, *101*
rivers 8, *9*, 54, 80, *81*
roads 76, 78, *79*, 90
root crops 40, 42
Romance languages 34
Rossosh 65
Rostov-na-Donu 36, *37*, *63*, 69, 71, *93*, 100, *101*
Rovno 65, 66, *71*, 93
Rozdol 65
rubber 66, 67
Rubtsovsk 63
Rudnyy 36, *49*, 57, *103*
Rumania 2, *83*
Rumanians 34
rural population *29*, 32, *33*, 90, 92
Russian Empire 26
Russian Federated Soviet Socialist Republic (RSFSR) 22, *23*, 24, *72*
Russian people 34, 35, 102
Rustavi *57*, 65, 67, 100, *101*
Ruthenia 26
Ryazan *37*, *63*, 65, 67, *71*, *91*
Rybinsk reservoir *63*, 88, *89*, *91*
rye 40, *41*, 44
Rzhev 69

Saami people *35*
Sadon *101*
Safonovo 67
Sagastyr 12, *13*
Sakar *105*
Sakhalin I, *51*, 52, *53*, 56, *99*
Salavat 64, 65
salts, mineral 48, *49*, 64, 66, *87*, *89*, 90, *91*, *93*, 94, *95*, *97*, *99*, *103*
Samantepe *105*
Samara 94
Samarkand *37*, 65, 69, 104, *105*
Samotlar *99*
San Francisco 26
Saransk 68, *69*, *71*, *91*
Sarany *49*, *97*
Sarapul *71*
Saratov *37*, *61*, 67, *71*, 94, *95*
Satka *57*, *97*
saw-milling *72*, *73*
saxaul 16
Sayak *49*, 102, *103*
Sayan mts 6, *19*, 44
Sayan-Shushenskoye 65, *98*, *99*
Scandinavia 80
semi-desert 16, *17*, *19*, 30, 44, *95*
Semipalatinsk 12, *37*, 69, *71*, *103*
Serov *49*, *57*, *71*, 96, *97*
serozem soils 14
Serpukhov 67, *69*, *91*

Sevan, lake 100
Sevastopol *37*, *93*
Severodonetsk 67, *93*
Severouralsk *49*, 96, *97*
Shadrinsk *69*, *71*
Shaim *99*
Shakhty *71*, *93*, *101*
shale, oil *53*, *87*, 88, *89*, 94
Shatlyk *105*
Shchelkovo *69*
Shchigry 65
Shebelinka 92, *93*
sheep 44
Shelekhov *59*, *99*
Shevchenko *103*
ships, shipbuilding 62, *63*, 88
Shorsu 65
Shubarkuluk *103*
Shuya 67, *69*
Shyaulyay *71*
Sibay *49*
Siberia 8, 10, 12, 14, 16, 24, 26, 28, 30, 32, *33*, 40, 42, 46, *47*, 48, 52, 54, 60, 62, 68, *72*, *73*, 78, 80, 86, 94, 98, *99*; *see also* East Siberia, West Siberia, Far East
Siberian platform 4, 48
silk 68, *69*
silver *49*
Simferopol *37*, 67, *71*, *93*
Skopin *61*
Slantsy *89*
Slav languages, peoples 26, 32, 34, *35*
Smolensk 18, *37*, 64, 65, *69*, *91*
snow 12
Sobinka *69*
Sochi *37*, *101*
soda, caustic 64
softwood *72*, *73*
soils 14, *15*
Soligorsk *49*, 65, 66, *89*
Solikamsk 48, *49*, 65, 66, 96, *97*
Sol-Iletsk *97*
Southern economic region 24, *25*, *29*, *87*, 92, *93*
South Osetian A.Ob. 22, *23*, *25*, 100, *101*
South-western economic region 24, *25*, 30, *87*, 92, *93*
South Siberian railway 76
Sovetskaya Gavan *63*, *99*
Sovetskiy *105*
Soviet of Nationalities 22, 24
Soviet Socialist Republics (SSRs). 22, *23*
Soviet of the Union 22
sown area 40, 42, 46
soya beans 42, *43*, 44
sphagnum 16
spruce 16
Sredneuralsk 65
Stalingrad 94; *see also* Volgograd
Stanovoy mts. 6, 26

Starobin 65
Staroutkinsk *57*, *97*
Stavropol 52, 56, *71*
Staryy Oskol *57*, 90, *91*
Stebnik 65
steel 48, 50, 56, *57*, 58, 60, 86, *87*, 88, *89*, 90, *91*, 92, *93*, 96, *97*, *99*, 100, *101*, 102, *103*, 104, *105*
Stepanakert *69*
Stepnyak *103*
steppe 12, 16, 18, *19*, 26, 28, 40, 44, 92, 100
Sterlitamak *49*, 65
structure, geological 4, 18
Stupino *69*
sub-Carpathian Ukraine (Ruthenia) 26, *27*, *93*
Suchan *99*
sugar 74, *75*
sugar beet 40, 42, *43*, 44, 45, 74, 100
Sukhona r. 8, 80, *81*, *89*
Sukhoy Log 96, *97*
Sukhumi *71*, *101*
sulphur 48, *49*, 64, 65, *87*, *95*, *93*, *103*, 104, *105*
sulphuric acid 64
Sumgait *57*, *59*, 65, 67, 100, *101*
Sumsar *105*
Sumy 65, *69*, *71*, *93*
Suna r. *89*
sunflower 44, *43*
Supreme Soviet 22
Surgut *99*
Surkhandarya oblast *105*
Sursk *69*
Svir r. *89*
Sverdlovsk 36, *37*, *57*, 60, *61*, 65, 67, *71*, *72*, *97*, *103*
Svetlogorsk 67, 88, *89*
Svetlyy *49*, *97*
Svobodnyy *63*
swamps 6, 40
Sweden 4, 26
Syktyvkar *63*, *71*, *89*
synthetic fibres 64, 66, 67, 104
synthetic textiles 68, *69*, 104
synthetic resins 64, 66, 67
synthetic rubber 64, 66, 67
Syrdarya oblast *105*
Syr Darya r. 8, *9*, *103*, 104, *105*
Syrzan *61*, 65, 94, *95*

Taboshar *105*
Tadzhikistan, Tadzhik SSR 22, *23*, 24, 104, *105*
Tadzhik people 34, *35*
Taganrog *37*, *57*, *61*, *71*, *93*, 100, *101*
Taldy-Kurgan *71*, *103*
Tallin *37*, *61*, 62, *63*, 65, *69*, *71*, 88, *89*
Talysh lowland 44
Talysh people *35*
Tambov *37*, 67, *71*, *91*
Tartu *63*, *71*

Tashauz *105*
Tashkent 30, 36, *37*, *61*, *63*, 65, 69, *71*, *105*
Tashtogul 57, *99*
Tat people *35*
Tatar ASSR 22, *23*, 94, *95*
Tatar people 22, 26, 34, *35*
Tavda r. *97*
Tavolzhan *103*
tayga 16, *17*, 18, *19*, 28, 30, 44, *72*
Taymyr (Dolgano-Nenets) A.Ok. 22, *23*, 28
Taymyr peninsula 2, *19*, *51*
Tbilisi *13*, 36, *37*, *61*, 62, *63*, 67, 69, *71*, *101*
tea 40, *43*, 44, *45*
Tedzhen *105*
Tekeli *103*
Telavi *69*
Temir Tau 57, *102*
temperatures 10, *11*, *12*, *13*
Terek r. 8, *101*
Ternopol *93*
Tertiary rocks 4, 5
Tetyukhe 49, *59*, *99*
textiles 68, *69*, 88, 90, 104
textile machinery 62
Tezhsar *100*, *101*
timber 16, *53*, *61*, *72*, *73*, 76, 77, 80, *81*, 94
tin 48, *49*, *87*, 98, *99*
tinned food 74
titanium 48, *87*, 93
Tkibuli *100*, *101*
Tkvarcheli *100*, *101*
tobacco *43*, 44, *75*, *83*, 100
Tobol r. *103*
Tokmak *93*
Toktogul *104*, *105*
Tolyatti (Togliatti) *37*, 62, *63*, 65, 66, 67, 94, *95*
Tom r. *99*
Tomsk *37*, *61*, *71*, *72*, *99*
Torzhok *71*
towns 36, *37*, *87*, *89*, *91*, *93*, *95*, *97*, *99*, *101*, *103*, 105
tractors 60, 62, 65, 88
trade *72*, 76, 78, 80, 82, *83*
transhumance 44
Trans-Aral railway 76
Transbaykalia 6, 8
Trans-Caspian railway 76
Transcaucasus, Transcaucasian economic region 10, *13*, *14*, 16, *24*, *25*, 28, *29*, 30, 32, 40, 42, 44, 46, *47*, 58, 64, 74, 78, 80, 86, *87*, 92, 94, 100, *101*; *see also* Armenia, Azerbaydzhan, Georgia
transport 76, *77*, *78*, *79*, 80, *81*, 90
transport engineering 62, *63*
Trans-Siberian railway 36, 50, *71*, 76, 98, *102*
Troitsk *71*
Tsaritsyn 94
Tselinograd *61*, *63*, *103*
Tuapse 65, *101*

Tula *37*, *57*, *61*, *63*, *71*, *91*
tundra 14, *15*, 16, *17*, 18, *19*, 30, 44, *72*
tung nuts 45, *100*
tungsten 49, *87*, 98, *99*, *100*, *101*, 104, *105*
Tunguska basin 4, 50, *51*
Tungus-Manchurian languages, peoples 34, *35*
Tura r. *97*
turbines 60
Turgay *103*
Turkestan 80; *see also* Central Asia, Kazakhstan
Turkestan-Siberian railway 76
Turkey 2, 6, 26
Turkic languages, peoples 34, *35*
Turkmenia, Turkmen SSR 22, *23*, 24, 42, 104, *105*
Turkmen people 34, *35*
Turtkul *12*, *13*
Turukhansk *12*, *13*
Tuvinian ASSR 22, *23*
Tuvinian people 34, *35*
Tuymazy *95*
Tyan Shan mts 6
tyres 66, 67
Tyrnyauz 49, *100*, *101*
Tyumen *33*, *37*, 52, *53*, *61*, *63*, 67, 69, *71*, *72*, 86, *99*
Tyuya Muyan *105*

Ubagan basin *51*
Uchaly *95*
Uchkyr *105*
Udmurt ASSR 22, *23*, *97*
Udmurt people 34, *35*
Udokan 27, 49
Ufa 36, *37*, 65, 67, 69, *71*, *95*
Ufa plateau *95*
Ufaley *97*
Uglich *91*
Ugro-Finnic languages, peoples 34, *35*
Ukhta 65, 88, *89*
Ukraine, Ukrainian SSR 4, 8, 10, *12*, *13*, 14, 22, *23*, 24, 28, 36, 46, 48, 50, 52, *53*, 54, 56, *57*, 60, 62, 68, *72*, 74, 76, 78, 80, 86, 92, *93*, 100
Ukrainian people 34, *35*, *102*
Ukrainian shield 4, 8
Ulan Ude *37*, *69*, *99*
Ulukhemsk basin *51*
Ulyanovsk *37*, 62, *63*, *71*, 94, *95*
Union Republics *see* SSRs
United Kingdom 2, 3, 54, 82
United States *see* America
Ural mts, economic region 4, 6, 18, *19*, 22, *25*, 26, *29*, 30, 34, 36, 46, 48, 50, *51*, 52, *53*, 54, 56, *57*, 58, 60, 62, 64, 66, 68, *72*, *73*, 76, 80, *81*, 86, *87*, 94, 96, *97*, 98, *102*, 104
Ural-Altaic languages, peoples 34, *35*
Ural-Kuznetsk combine 98
Ural r. 8, *9*, *97*, *103*
Uralsk *71*, *103*
uranium *87*, *105*

urbanisation 18, *29*, 30, *33*, 36, *37*, 90, 92
Urengoy *99*
Urgal *99*
Urta-Bulak *105*
Urup 49, *101*
Usolye-Sibirskoye 48, *49*, 60, *61*, 66, 67, *71*, *99*
Ussuri r. 8, 26
Ussuriysk *61*, *71*
Ust-Ilim 54, 98, *99*
Ust-Kamenogorsk *37*, 58, *59*, *61*, 62, 65
Ust Katav *63*
Ust Kut *63*
Ust Leninogorsk *103*
Usk Orda Buryat A.Ok. *23*
Ust Tsilma *71*
Ust Yurt plateau 6
Uvarovo 65
Uzbekistan, Uzbek SSR 22, *23*, 24, 32, 42, 58, 104, *105*
Uzbek people 34, *35*
Uzhgorod *71*
Uzlovaya *61*

Vaksh r. *105*
Valday hills 8, 18
Valmiera 67
vanadium 48, *49*
Varsob r. *105*
Vartsilya *57*, *89*
vegetables 40, 44, 46, *47*, 75
vegetable oils 74
vegetation 14, 16, *17*, 18, *19*, *72*, 74
Velikiye Luki *71*
Ventspils *89*
Verkhniy Ufaley *57*, *59*, *97*
Verkhnyaya Pyshma *59*
Verkhnyaya Salda *57*, *61*, *97*
Verkhnyaya Sinyachika *57*
Verkhoyansk *12*, *13*
Vichuga *69*
Vilnyus *37*, *63*, *71*, 88, *89*
Vilyuy r. 8
vines *43*, 44, *100*
Vinnitsa *41*, 65, 66, *93*
virgin lands 44, *102*
Vitebsk *41*, *69*, *71*, *93*
Vitim r. 8
Vladimir *37*, 67, *69*, *71*, *91*
Vladivostok 10, *13*, *37*, *71*, *99*
Volga economic region 14, 24, *25*, *29*, 40, 42, 48, 62, 68, *72*, *73*, 86, *87*, 94, *95*, *101*
Volga r. 6, *9*, 18, 22, 26, 34, 36, 44, 46, 52, 54, 55, 68, *72*, 74, 80, *81*, 90, 94, *95*, *101*
Volga-Baltic waterway 80, *81*
Volga-Don canal 80, *81*, *95*
Volga heights, upland 6, *95*
Volga-Ural oilfield 52, *53*, 64, 94, *95*, 96, *97*

116

Volga–Vyatka economic region 24, 25, 29, 40, 42, 46, 72, 73, 86, 87, 90, 91, 92
Volgograd 36, 37, 54, 58, 59, 62, 63, 65, 67, 71, 94, 95
Volkhov r. 8, 58, 59, 65, 88, 89
Vologda 69, 71, 72, 89
Volyn oblast 93
Volyn upland 6, 8
Volzhskiy 67, 94, 95
Vorkuta 88, 89
Voronezh 37, 57, 61, 63, 69, 71, 93
Voronya r. 88, 89
Voroshilovgrad 28, 37, 57, 61, 63, 69, 71, 92, 93
Voskresensk 65
Votkinsk 97
Vyazniki 69
Vyborg 71
Vychegda, r. 8, 89
Vyg, r. 88, 89
Vyksa 57, 91
Vyshniy Volochek 69
Vyosokovsk 69

Warsaw, Duchy of 26, 27
waterways 80, 81, 90
West Siberian economic region 14, 24, 25, 28, 29, 30, 33, 34, 40, 42, 46, 52, 53, 58, 62, 76, 87, 88, 90, 96, 98, 99
West Siberian lowland 4, 6, 8, 16, 18, 19, 26, 52, 53, 98, 99
wheat 40, 41, 42, 44, 100
White Sea 60, 80, 81, 94
White Sea–Baltic canal 80

willow 16
winds 10, 11
wine 74, 75
wooded steppe 16, 17, 18, 28, 40, 42, 44, 92, 95
wooded tundra 18, 19
woodworking 72, 73
wool 68, 69
World War I 26, 32, 94
World War II 26, 32, 34, 50, 56, 58, 88, 94, 98

Yakhroma 69
Yaklovo 57
Yakut ASSR 22, 23, 30, 48, 99
Yakut basin 18, 19, 51, 52
Yakut people 34, 35
Yakutsk 12, 13, 71, 98, 99
Yalta 13, 93
Yamalo-Nenets A.Ok. 22, 23, 33
Yaroslavl 37, 61, 63, 65, 67, 68, 69, 91
Yartsevo 69
Yasnogorsk 61
Yavan 66
Yefremov 67
Yegorevsk 69, 71
Yelets 71
Yelgava 69
Yenikayevo 57, 95
Yenisey r. 9, 55, 81, 99
Yerevan 36, 37, 59, 61, 66, 67, 69, 71, 100, 101
Yermak 57, 102, 103

Yermolino 69
Yevreysk (Jewish) A.Ob. 22, 23, 99
Yoshkar-Ola 71, 91
Yuryev Polskiy 69
Yuzha 69
Yuzhno-Kurilsk I. 2
Yuzhno-Sakhalinsk 3, 71
Yuzhno-Sukhokumsk 101

Zagorsk 91
Zakarpatny oblast 93
Zakamensk 49
Zapolyarnoye 49, 99
Zaporozhye 58, 71, 92, 93
Zaterechnyy 101
Zeravshan 105
Zestafoni 57, 100, 101
Zeya r. 99
Zhanatas 65, 66, 103
Zhdanov 37, 56, 57, 60, 61, 65, 93
Zheleznogorsk 56, 57, 98, 99
Zhigulevsk 94, 95
Zhitomir 37, 71, 93
Zhodino 62, 63
zinc 48, 49, 58, 59, 87, 98, 99, 100, 101, 102, 103, 104, 105
zirconium 93
Zlatoust 57, 71, 97
Zyryanka basin 51
Zyryanovsk 103